FACTORIZATION THEORY OF MEROMORPHIC FUNCTIONS

PURE AND APPLIED MATHEMATICS

A Program of Monographs, Textbooks, and Lecture Notes

LECTURE NOTES
IN PURE AND APPLIED MATHEMATICS

1. *N. Jacobson*, Exceptional Lie Algebras
2. *L.-Å. Lindahl and F. Poulsen*, Thin Sets in Harmonic Analysis
3. *I. Satake*, Classification Theory of Semi-Simple Algebraic Groups
4. *F. Hirzebruch, W. D. Newmann, and S. S. Koh*, Differentiable Manifolds and Quadratic Forms (out of print)
5. *I. Chavel*, Riemannian Symmetric Spaces of Rank One (out of print)
6. *R. B. Burckel*, Characterization of C(X) Among Its Subalgebras
7. *B. R. McDonald, A. R. Magid, and K. C. Smith*, Ring Theory: Proceedings of the Oklahoma Conference
8. *Y.-T. Siu*, Techniques of Extension of Analytic Objects
9. *S. R. Caradus, W. E. Pfaffenberger, and B. Yood*, Calkin Algebras and Algebras of Operators on Banach Spaces
10. *E. O. Roxin, P.-T. Liu, and R. L. Sternberg*, Differential Games and Control Theory
11. *M. Orzech and C. Small*, The Brauer Group of Commutative Rings
12. *S. Thomeier*, Topology and Its Applications
13. *J. M. López and K. A. Ross*, Sidon Sets
14. *W. W. Comfort and S. Negrepontis*, Continuous Pseudometrics
15. *K. McKennon and J. M. Robertson*, Locally Convex Spaces
16. *M. Carmeli and S. Malin*, Representations of the Rotation and Lorentz Groups: An Introduction
17. *G. B. Seligman*, Rational Methods in Lie Algebras
18. *D. G. de Figueiredo*, Functional Analysis: Proceedings of the Brazilian Mathematical Society Symposium
19. *L. Cesari, R. Kannan, and J. D. Schuur*, Nonlinear Functional Analysis and Differential Equations: Proceedings of the Michigan State University Conference
20. *J. J. Schäffer*, Geometry of Spheres in Normed Spaces
21. *K. Yano and M. Kon*, Anti-Invariant Submanifolds
22. *W. V. Vasconcelos*, The Rings of Dimension Two
23. *R. E. Chandler*, Hausdorff Compactifications
24. *S. P. Franklin and B. V. S. Thomas*, Topology: Proceedings of the Memphis State University Conference
25. *S. K. Jain*, Ring Theory: Proceedings of the Ohio University Conference
26. *B. R. McDonald and R. A. Morris*, Ring Theory II: Proceedings of the Second Oklahoma Conference
27. *R. B. Mura and A. Rhemtulla*, Orderable Groups
28. *J. R. Graef*, Stability of Dynamical Systems: Theory and Applications
29. *H.-C. Wang*, Homogeneous Banach Algebras
30. *E. O. Roxin, P.-T. Liu, and R. L. Sternberg*, Differential Games and Control Theory II
31. *R. D. Porter*, Introduction to Fibre Bundles
32. *M. Altman*, Contractors and Contractor Directions Theory and Applications
33. *J. S. Golan*, Decomposition and Dimension in Module Categories
34. *G. Fairweather*, Finite Element Galerkin Methods for Differential Equations
35. *J. D. Sally*, Numbers of Generators of Ideals in Local Rings
36. *S. S. Miller*, Complex Analysis: Proceedings of the S.U.N.Y. Brockport Conference
37. *R. Gordon*, Representation Theory of Algebras: Proceedings of the Philadelphia Conference
38. *M. Goto and F. D. Grosshans*, Semisimple Lie Algebras
39. *A. I. Arruda, N. C. A. da Costa, and R. Chuaqui*, Mathematical Logic: Proceedings of the First Brazilian Conference

40. *F. Van Oystaeyen*, Ring Theory: Proceedings of the 1977 Antwerp Conference
41. *F. Van Oystaeyen and A. Verschoren*, Reflectors and Localization: Application to Sheaf Theory
42. *M. Satyanarayana*, Positively Ordered Semigroups
43. *D. L. Russell*, Mathematics of Finite-Dimensional Control Systems
44. *P.-T. Liu and E. Roxin*, Differential Games and Control Theory III: Proceedings of the Third Kingston Conference, Part A
45. *A. Geramita and J. Seberry*, Orthogonal Designs: Quadratic Forms and Hadamard Matrices
46. *J. Cigler, V. Losert, and P. Michor*, Banach Modules and Functors on Categories of Banach Spaces
47. *P.-T. Liu and J. G. Sutinen*, Control Theory in Mathematical Economics: Proceedings of the Third Kingston Conference, Part B
48. *C. Byrnes*, Partial Differential Equations and Geometry
49. *G. Klambauer*, Problems and Propositions in Analysis
50. *J. Knopfmacher*, Analytic Arithmetic of Algebraic Function Fields
51. *F. Van Oystaeyen*, Ring Theory: Proceedings of the 1978 Antwerp Conference
52. *B. Kedem*, Binary Time Series
53. *J. Barros-Neto and R. A. Artino*, Hypoelliptic Boundary-Value Problems
54. *R. L. Sternberg, A. J. Kalinowski, and J. S. Papadakis*, Nonlinear Partial Differential Equations in Engineering and Applied Science
55. *B. R. McDonald*, Ring Theory and Algebra III: Proceedings of The Third Oklahoma Conference
56. *J. S. Golan*, Structure Sheaves over a Noncommutative Ring
57. *T. V. Narayana, J. G. Williams, and R. M. Mathsen*, Combinatorics, Representation Theory and Statistical Methods in Groups: YOUNG DAY Proceedings
58. *T. A. Burton*, Modeling and Differential Equations in Biology
59. *K. H. Kim and F. W. Roush*, Introduction to Mathematical Consensus Theory
60. *J. Banas and K. Goebel*, Measures of Noncompactness in Banach Spaces
61. *O. A. Nielsen*, Direct Integral Theory
62. *J. E. Smith, G. O. Kenny, and R. N. Ball*, Ordered Groups: Proceedings of the Boise State Conference
63. *J. Cronin*, Mathematics of Cell Electrophysiology
64. *J. W. Brewer*, Power Series Over Commutative Rings
65. *P. K. Kamthan and M. Gupta*, Sequence Spaces and Series
66. *T. G. McLaughlin*, Regressive Sets and the Theory of Isols
67. *T. L. Herdman, S. M. Rankin, III, and H. W. Stech*, Integral and Functional Differential Equations
68. *R. Draper*, Commutative Algebra: Analytic Methods
69. *W. G. McKay and J. Patera*, Tables of Dimensions, Indices, and Branching Rules for Representations of Simple Lie Algebras
70. *R. L. Devaney and Z. H. Nitecki*, Classical Mechanics and Dynamical Systems
71. *J. Van Geel*, Places and Valuations in Noncommutative Ring Theory
72. *C. Faith*, Injective Modules and Injective Quotient Rings
73. *A. Fiacco*, Mathematical Programming with Data Perturbations I
74. *P. Schultz, C. Praeger, and R. Sullivan*, Algebraic Structures and Applications: Proceedings of the First Western Australian Conference on Algebra
75. *L. Bican, T. Kepka, and P. Němec*, Rings, Modules, and Preradicals
76. *D. C. Kay and M. Breen*, Convexity and Related Combinatorial Geometry: Proceedings of the Second University of Oklahoma Conference
77. *P. Fletcher and W. F. Lindgren*, Quasi-Uniform Spaces
78. *C.-C. Yang*, Factorization Theory of Meromorphic Functions

Other Volumes in Preparation

FACTORIZATION THEORY OF MEROMORPHIC FUNCTIONS AND RELATED TOPICS

edited by

Chung-Chun Yang

Naval Research Laboratory
Washington, D.C.

MARCEL DEKKER, INC.   New York and Basel

Library of Congress Cataloging in Publication Data
Main entry under title:

Factorization theory of meromorphic functions, and
related topics.

(Lecture notes in pure and applied mathematics ;
v. 78)
Includes index.
1. Functions, Meromorphic--Addresses, essays,
lectures. 2. Factorization (Mathematics)--Addresses,
essays, lectures. I. Yang, Chung-Chun, 1942- .
II. Series.
QA331.F28 1982 515.9'82 82-9757
ISBN 0-8247-1834-8 AACR2

MARCEL DEKKER, INC.

270 Madison Avenue, New York, New York 10016

Current printing (last digit):

10 9 8 7 6 5 4 3 2 1

PRINTED IN THE UNITED STATES OF AMERICA

In memory of my father;

to my mother and my wife--Chwang-Chia

PREFACE

Factorization theory of meromorphic functions is essentially a study of the ways in which a given entire or meromorphic function can be expressed as a composition of two or more entire or meromorphic functions. More specifically, let F(z) be a meromorphic function such that $F = f_1 o f_2 \cdots o f_n$, where each f_i is entire or meromorphic, then we have a factorization of F, and f_i are called factors. F is called prime (pseudo-prime) if every factorization of the form F = fog = f(g) implies that either f or g is linear (polynomial). J. F. Ritt was the first one who dealt with factorization theory for polynomials and obtained fairly complete results in this aspect. The study of factorization for transcendental entire function was first introduced by P. C. Rosenbloom in a paper in which he mainly investigated the fix-points of iteration of transcendental entire functions. There he proved that for every transcendental entire function f, its second iterate f(f) must have fix-points. That is, the identity $f(f) = z + e^{\alpha(z)}$ is impossible for any entire function $\alpha(z)$. In the same paper, he stated without proof that $e^z + z$ is prime.

Later on, F. Gross and I. N. Baker provided a proof of this assertion and generalized the result to various classes of entire and meromorphic functions. Shortly after this, M. Ozawa, C. C. Yang, R. Goldstein, A. A. Gol'dberg, G. S. Prokopovich, E. Mues, K. Niino, H. Urabe and others have joined in the pursuit of the subject of factorization theory in several directions such as primeness, permutability, and unique factorizability. The basic tools in the investigation of this subject, thus far, are mainly classical function theory, Nevanlinna's value-distribution theory, and some of the extensions of Nevanlinna's theory to algebroid functions. Clearly, in order to pursue this subject further, more detailed studies on the relationship between the growth rates of a meromorphic function and its possible factors are needed. Also needed is the study of the relationship between the growth rate of a meromorphic function and the geometric distribution of the preimage sets of some finite or infinite range set of the function. All in all, there are numerous, interesting questions which can be formed and still need to be answered. Frankly speaking, the development of this subject is still in its infancy.

Therefore, the purpose of the present volume is twofold. First, it is hoped that this volume will bring attention and spread interest among mathematicians to this area. Second, to introduce some newly obtained results which are in, or related to, this subject.

Finally, the editor would like to express thanks to the authors of this volume and to the editorial staff of marcel dekker, inc. who worked hard and enthusiastically to make possible the publication of this volume.

Chung-Chun Yang

CONTENTS

CONTRIBUTORS

I. N. Baker
Department of Mathematics, Imperial College, London, England

F. Gross
University of Maryland - Baltimore County, Baltimore, Maryland; and Naval Research Laboratory, Washington, D. C.

Tadashi Kobayashi
Chiba Keiai University, Chiba, Japan

L. S. O. Liverpool[1]
Department of Mathematics, Fourah Bay College, University of Sierra Leone, Freetown, Sierra Leone

Kiyoshi Niino
Faculty of Technology, Kanazawa University, Kanazawa, Japan

Charles F. Osgood
Naval Research Laboratory, Washington, D. C.

Mitsuru Ozawa
Department of Mathematics, Tokyo Institute of Technology, Tokyo, Japan

Norbert Steinmetz
Universitat Karlsruhe, Karlsruhe, Federal Republic of Germany

Chen-Han Sung[2]
Department of Mathematics, Notre Dame University, Notre Dame, Indiana

Hironobu Urabe
Kyoto University of Education, Kyoto, Japan

Chung-Chun Yang
Naval Research Laboratory, Washington, D. C.

[1]*Current affiliation* - Department of Mathematics, University of Jos, Jos, Nigeria

[2]*Current affiliation* - Department of Mathematics, University of California, Santa Barbara, California

ENTIRE FUNCTIONS WHOSE A-POINTS LIE ON SYSTEMS OF LINES

I. N. Baker

Department of Mathematics
Imperial College
London, England

1. INTRODUCTION

If f is a transcendental entire function such that for each complex w the roots of f(z) = w lie on a line $\ell(w)$ of the complex plane, then there are constants a, b, c such that $f(z) = ae^{bz} + c$ [1].

Recently this result has been extended in different ways by T. Kobayashi [5, 6] and M. Ozawa [8]. We may replace the line $\ell(w)$ by a system of say at most k lines. Taking k = 2, Ozawa [8] showed

THEOREM A. Let f be transcendental entire, real for real z and of finite order. Assume that for every real w satisfying $|w| > |w_0|$ the roots of f(z) = w lie on two lines $\ell(w)$, $\ell'(w)$ parallel to the real axis and that for every real w satisfying $|w| < |w_0|$ the roots of f(z) = w lie on two straight lines $\ell(w)$, $\ell'(w)$. Then there are real constants a, b, c, d such that $f(z) = a \cos(bz+c) + d$, $ab \neq 0$.

The assumptions in theorem A seem more elaborate than necessary. In fact one can prove

THEOREM 1. Let f be transcendental entire, k a positive integer and G an open plane set such that for every w in G there is a system of $j(w) \leq k$ straight lines which contain all the solutions z of $f(z) = w$. Suppose moreover that for some w in G the system of lines cannot contain fewer than k lines.

Then all the lines (for all w) may be taken to be parallel, and there is a constant $A \neq 0$ such that $f(z) = R(e^{Az})$, where R is a rational function of order k.

In the case $k = 2$ our results yield the following improvement of theorem A:

THEOREM 2. Let f be transcendental entire and real on the real axis. Assume that for some open set of w-values the solutions of $f(z) = w$ lie on two lines $\ell(w)$, $\ell'(w)$ and that for at least one of these values two different lines $\ell(w)$, $\ell'(w)$ are needed. Then there are real constants a, b, c, d such that $ab \neq 0$ and $f(z) = a\cos(bz+c) + d$.

One can also seek to reduce further the size of the set of w for which one supposes the w-points of f to be distributed on lines. In the case $k = 1$, T. Kobayashi [5, 6] has proved

THEOREM B. Let f be transcendental entire and w_1, w_2, w_3 three distinct values which are not collinear in the w-plane and are such that the roots of $f(z) = w_i$ lie on a line ℓ_i for $i = 1, 2, 3$. Then there is a non-zero constant A and a polynomial P of degree at most two such that

$$f(z) = P(e^{Az})$$

In fact transcendental functions of the above form do have the property assumed in the theorem.

We give a proof of theorem B which is somewhat shorter than that in Kobayashi's papers, although the approach is similar. It would be interesting to have a result like that of theorem B in the case when $k > 1$, but so far as I know nothing of this kind has been proved.

2. PROOF OF THEOREMS 1 AND 2

Denote the spherical derivative of an entire or meromorphic function g by

$$\rho(g(z)) = \frac{|g'(z)|}{1 + |g(z)|^2}$$

and

$$\mu(r,g) = \sup_{|z|=r} \rho(g(z))$$

Further, put $M(r,g) = \text{Max}|g(z)|$ for $|z| = r$.

We quote from Clunie and Hayman [3, theorem 3] the

LEMMA 1. If $g(z)$ is entire and σ a constant such that $-1 < \sigma < \infty$ and there are constants $K > 0$ and r_0 such that

$$\mu(r,g) \le Kr^{\sigma}, \qquad r > r_0$$

then for all sufficiently large r

$$\log M(r,g) < \frac{A_1 K}{\sigma+1} r^{\sigma+1}$$

where $A_1 = 25e \log 2$.

Let $h(r)$ be a positive function such that $h(r) = o(r)$ as $r \to \infty$. From Lehto [7] we have

LEMMA 2. Let $g(z)$ be meromorphic in $R < |z| < \infty$. If for a sequence (z_k), $\lim|z_k| = \infty$ and

$$\lim_k h(|z_k|)\rho(g(z_k)) = \infty$$

then Picard's theorem holds for g in the union of any infinite subsequence of the discs

$$C_k = \{z : |z-z_k| < \varepsilon h(|z_k|)\}$$

for each $\varepsilon > 0$.

From these results follows

LEMMA 3. Let f be transcendental entire and α_i, $1 \le i \le 3$, three different complex numbers such that the roots of $f(z) = \alpha_i$ lie on finitely many lines whose union we denote by L_i. Suppose further that the lines of L_i are all different from those of L_j if $i \neq j$. Then f is at most of order 1 and finite exponential type, i.e., there is a constant A such that for all large r

$$\log M(r,f) < Ar.$$

For suppose not. Then by lemma 1 with $\sigma = 0$ there exists a sequence z_k such that $z_k \to \infty$ and $\rho(f(z_k)) \to \infty$. Choose $\varepsilon > 0$ so that 4ε is less than the separation of every pair of parallel lines in $\bigcup_i L_i$. By lemma 2 Picard's theorem holds in any subsequence of the discs $C_k = \{z : |z-z_k| < \varepsilon\}$. For a fixed sequence of the C_k, f takes at least one of the α_i, say α_1, infinitely often, and we may suppose α_1 to be taken in every C_k if we relabel the discs. Thus each z_k must have distance at most ε from L_1, and by going over to a subsequence of (z_k) we may suppose the points z_k have distance at most ε from a fixed line ℓ_1 in L_1, while $|z_k| \to \infty$.

Observing that f must take at least one of the remaining values α_2 and α_3 (say α_2) infinitely often in the discs C_k, it can be seen that a subsequence of (z_k) has distance at most ε from a fixed line ℓ_2 of L_2. Since $\ell_1 \neq \ell_2$, this can occur only if ℓ_1 and ℓ_2 are parallel with separation at most 2ε. This contradicts the choice of ε, and the lemma is proved.

Proof of Theorem 1. For an entire function f and a straight line ℓ, the image $f(\ell)$ has zero plane measure. If for each w in some open set G the roots of $f(z) = w$ lie on a finite system of lines $L(w)$, we can find three values of w, say α_1, α_2, and α_3, for which the lines of $L(\alpha_i)$ are different from those of $L(\alpha_j)$ if $i \neq j$. By lemma 3 the function f is at most of order one and finite exponential type.

Now take a value $b \in G$ such that the roots of $f(z) = b$ lie on the maximal number k of lines and not on fewer. There is therefore a finite subset, say $S(b)$, of the set of roots of $f(z) = b$ which does not lie on any $(k-1)$ straight lines. For b' near b there is a set $S(b')$ of roots of $f(z) = b'$ whose elements are near those of $S(b)$, and if b' is sufficiently close to b, the points of $S(b')$ cannot lie on any $(k-1)$ straight lines. Thus by replacing b by a neighbouring value if necessary we can suppose that b is not Picard exceptional and that for all values w in a neighbourhood of b there must be k lines in $L(w)$. We note also from the order of f that the singularities of $f^{-1}(w)$ consist of an at most countable set of algebraic singularities over points $f(\xi)$ such that $f'(\xi) = 0$, together with at most two finite transcendental singularities corresponding to asymptotic values of f. Thus singularities of f^{-1} lie over at most countably many base points and we may suppose b has been chosen different from all these base points.

Among the b-points of f there is an infinite subset which lie on one of the lines, say $\ell_1(b)$, of $L(b)$, and there is some finite subset say $T(b)$

of b-points which do not lie on $\ell_1(b)$ and which are contained in (k-1) lines but do not lie on any (k-2) straight lines. Since all branches of $f^{-1}(w)$ are analytic at w = b, we have for any b' sufficiently near b a set of b' points T(b') near those of T(b) and contained in (k-1) lines but not fewer. If $\phi(b')$ denotes a branch of f^{-1} such that $\phi(b) \in \ell_1(b)$ and if $\phi(b')$ is collinear with two fixed branches in T(b') for a sequence of b' with limit b, then $\phi(b)$ is collinear with two members of T(b). This can happen at most for finitely many branches.

We choose any three different branches ϕ_i, $1 \leq i \leq 3$, such that $\phi_i(b) \in \ell_1(b)$ and $\phi_i(b')$ is not collinear with two members of T(b') for b' near b. Then $\phi_i(b')$ lie on the k lines of L(b') and since (k-1) lines are needed to cover T(b'), it follows that $\phi_i(b')$ are collinear. Thus the ratio $(\phi_3-\phi_2)/(\phi_2-\phi_1)$ is real and analytic for b' near b and so is a real constant. Thus ϕ_1, ϕ_2, and ϕ_3 remain collinear not only for b' near b but for all b' so long as we continue along a path on which each of the ϕ_i remains analytic. This is true of all branches ϕ such that $\phi(b) \in \ell_1(b)$, with perhaps finitely many exceptions. Discard the exceptions and, starting from a fixed branch $\phi_1(b) \in \ell_2(b)$, label the (infinitely many) branches $\phi_1, \phi_2, \phi_3, \ldots$ so that $\phi_1(b), \phi_2(b), \phi_3(b), \ldots$ lie in that order on $\ell_1(b)$. If there are further branches on the opposite side of $\phi_1(b)$ from $\phi_2(b)$, these are labelled $\phi_0, \phi_{-1}, \ldots$.

Take a path γ from b to b in the w-plane such that $\phi_1(b)$ continues analytically along γ to $\phi_2(b)$. We can assume that γ has been chosen so as to avoid all the (at most countable) set of singularities of f^{-1}. Then each branch $\phi_j(b)$ continues along γ to a new branch $\phi_{\alpha(j)}(b)$. Since each of the expressions $(\phi_k-\phi_j)/(\phi_j-\phi_i)$ is analytic and so remains real and constant on γ, we see that $j \to \alpha(j)$ is a permutation which preserves relations of separation among the $\phi_i(b)$. Since $\alpha(1) = 2$, we have either $\alpha(2) = 3$ or $\alpha(2) = 1$ and the only possibilities are

(i) $\alpha(j) = j+1$

or (ii) $\alpha(j) = 3-j$

In both cases it is clear that j must in fact run from $-\infty$ to ∞.

If we take any member η of $f^{-1}(b)$ not lying on $\ell_1(b)$ and continue all the branches ϕ_i around a curve γ' on which they remain analytic and on which ϕ_1 continues to ϕ' such that $\phi'(b) = \eta$, the branches ϕ_i remain collinear, and so we see that there is a line $\ell_2(b)$ which passes through η and is thus different from $\ell_1(b)$ while containing an infinite subset of $f^{-1}(b)$.

Continuing the argument, we see that L(b) consists of k lines, each of which contains an infinite subset of $f^{-1}(b)$.

Since f has order one and exponential type, it follows that the counting function n(r,b) of the b-points of f in $|z| \leq r$ satisfies $n(r,b) < Kr$ for some constant K.

We now make the

CLAIM. The k lines of L(b) are parallel and f is periodic, the period being the distance between adjacent b-points on any line of L(b).

Proof of the Claim in Case (i). Analytic continuation of $(\phi_3-\phi_2)/(\phi_2-\phi_1)$ around γ shows that at w = b

$$(\phi_3-\phi_2)/(\phi_2-\phi_1) = (\phi_4-\phi_3)/(\phi_3-\phi_2)$$

and hence for all integers n we have at w = b

$$(\phi_{n+1}-\phi_n)/(\phi_n-\phi_{n-1}) = \lambda$$

where λ is a real positive constant. For w = b the assumption that $\lambda \neq 1$ leads to

$$\phi_n = \phi_1 + (\phi_2-\phi_1)\left(\frac{\lambda^{n-1}-1}{\lambda}\right), \qquad n = 0, \pm 1, \pm 2, \ldots .$$

which makes $\phi_n(b)$ have a finite limit point. Thus $\lambda = 1$ and $\phi_n-\phi_{n-1} = \phi_2-\phi_1$ for all n.

Put $\psi(b) = \phi_2(b)-\phi_1(b)$. If $|\psi(b)| = \delta$, then our estimate $n(r,b) < Kr$ shows that $\delta \geq 2/K$. Continuation of the branches ϕ_i along any path which avoids the singularities of f^{-1} keeps the branches collinear and ordered in the same way. Thus continuation around a closed path to b leads to a new value of $\psi(b)$ which is still given by $\phi'(b) - \phi''(b)$ where ϕ' and ϕ'' are two adjacent branches on one of the lines of L(b). On this line the branches are equally spaced with a spacing at least 2/K, as shown. Thus $\chi = 1/\psi$ is a function which has a finite number $p \leq k$ of branches under analytic continuation, the number necessarily being the same for different values of b. The algebraic singularities of a given branch ϕ_1 or ϕ_2 are isolated and thus χ has only isolated singularities at such points. The elementary symmetric functions of the branches $\chi_1, \ldots, \chi_p$ are single-valued and bounded near these isolated algebraic singularities, and hence these singularities are removable. The same is then also true of the (at most two) transcendental singularities of χ arising from transcendental singularities of f^{-1}.

By Liouville's theorem the elementary symmetric functions of the χ_i are constant and the χ_i are thus roots of an equation $\chi^p + A_{p-1}\chi^{p-1} \ldots + A_0 = 0$ in which the coefficients A_q are constants. Thus the χ_i derived by continuation from $1/\{\phi_2(b)-\phi_1(b)\}$ is a constant for all b. Thus $\phi_2(b) = \phi_1(b) + c$ for all b and f is periodic with period c.

Hence the k-lines of L(b) are all parallel, and the spacing between adjacent points on any line of L(b) is $|c|$.

Proof of the Claim in Case (ii). Consider a path τ from b to b on which $\phi_1(b)$ continues to $\phi_3(b)$. Continuation of the branches $\phi_i(b)$ around τ gives a permutation preserving their order on $\ell_1(b)$ in which $\phi_i \to \phi_{\beta(i)} = 3$ so that either $\beta(j) = j+2$ or $\beta(j) = 4-j$.

If $\beta(j) = 4-j$, we see that continuation along $\gamma^{-1}\tau$ takes ϕ_{3-j} into ϕ_{4-j}, which is the same permutation of branches as in case (i), so that the claim is proved.

If $\beta(j) = j+2$, we examine the analytic continuation of $\{\phi_4(b)-\phi_2(b)\}/\{\phi_2(b)-\phi_0(b)\}$ and find that

$$\{\phi_{2j+2}(b) - \phi_{2j}(b)\}/\{\phi_{2j}(b) - \phi_{2j-2}(b)\} = \lambda$$

where λ is a real positive constant, which can only be 1. Similarly $\{\phi_{2j+3}(b) - \phi_{2j+1}(b)\}/\{\phi_{2j+1}(b) - \phi_{2j-1}(b)\} = 1$. In order to keep the correct ordering of branches as $j \to \infty$, it is necessary to have $\phi_2(b)-\phi_0(b) = \phi_3(b)-\phi_1(b)$ and so all $\phi_{j+2}(b)-\phi_j(b)$ are equal.

Now apply the argument of the preceding case to $\psi(b) = \phi_3(b)-\phi_1(b)$, which is at most k-valued and satisfies $|\psi(b)| \geq 2/K$. It follows that f is periodic with a period $c = \phi_3(b)-\phi_1(b)$, and that all the lines $L_1(b)$ are parallel.

Thus for b' near b we have $\phi_3(b') = \phi_1(b') + c$, while $\phi_1(b')$, $\phi_2(b')$, and $\phi_3(b')$ are collinear. Hence

$$\{\phi_2(b') - \phi_1(b')\}/\{\phi_3(b') - \phi_1(b')\} = \mu$$

where μ is a constant independent of b' and $0 < \mu < 1$. This shows that $\phi_2(b') = \phi_1(b') + \mu c$ so that μc is a period of f. Since there are no b-points between $\phi_1(b)$ and $\phi_3(b)$ except $\phi_2(b)$, this is possible only if $\mu = \frac{1}{2}$. Thus the spacing of the ϕ_i is equal, namely $\frac{c}{2}$, and $\frac{c}{2}$ is the smallest period of f. The claim is now established in all cases.

To complete the proof of theorem 1, denote the period $\phi_1(b) - \phi_0(b)$ of f by λ and put $A = (2\pi i)/\lambda$. Then e^{Az} has the same value at each of the

$\phi_i(b)$, i.e., at all the b-points on $\ell_1(b)$. Making a similar observation for each of the k-lines of L(b), we see that there are k constants β_j such that

$$f(z) - b = g(z) \prod_{j=1}^{k} (e^{Az} - \beta_j)$$

where g must have the form Be^{Lz} and further be periodic with period λ. Thus $L = pA$ for some integer p. Thus $f(z) = R(e^{Az})$ where R is the rational function

$$R(w) = b + Bw^P \prod_{j=1}^{k} (w - \beta_j)$$

There is only a finite number of values c such that R(w) - c has multiple roots, Thus we can take c in place of b in the set G such that the c-points of f lie on exactly k-lines, that the "claim" applies to L(c), and such that R(w) = c has all its roots different. If $p > 0$ or $< -k$, there must be more than k-solutions w_i of R(w) - c; and if the solutions of $e^{Az} = w_i$, all i, are to lie on k lines, the spacing on one of the lines must be closer than $(2\pi i)/A = \lambda$.

Since $-k \leq p \leq 0$, the function R is rational of order k, as claimed.

Proof of Theorem 2. Put k = 2 in the proof of theorem 1, Since f is real on the real axis, the period must be real and A = ib, b real, $b \neq 0$. The form of f must be

$$f(z) = e^{pibz}(ae^{2ibz} + ce^{ibz} + d)$$

p = 0, -1, or -2.

Suppose p = 0. Then f(0) and $f^{(n)}(0)$, $n > 0$, are real so that a + c + d and $a(2ib)^n + c(ib)^n$ are real. n = 0, 2, 4 show a, c, and d must be all real while n = 1, 3 then shows this is possible only if a = c = 0 which makes f constant, against assumption.

The case p = -2 is exactly similar and corresponds to p = 0 by changing the sign of b.

Hence p = -1 and

$$f(z) = ae^{ibz} + c + de^{-ibz}$$

Consideration of the real quantities f(0), f'(0), f"(0) shows that c, a+d, and i(a-d) are real so that a and d are conjugates. Thus $a = \rho e^{i\alpha}$, $d - \rho e^{-i\alpha}$, ρ, α real and

$$f(z) = c + 2\rho \cos(bz+\alpha)$$

which is of the form asserted. Conversely, any such function satisfies the conditions of the theorem.

3. PROOF OF THEOREM B

We give a rather shorter proof than the original of Kobayashi's theorem B. We collect some preliminary results:

LEMMA 4. Let $g(z)$ be analytic in $H : \operatorname{Im} z > 0$ and omit the values 0 and 1 in H. Then there exists a constant $K = K(g)$ such that

$$\log|g(re^{i\theta})| < Kr/(\sin\theta) \qquad \text{for } r > 1,\ 0 < \theta < \pi \tag{1}$$

Proof. $z = \phi(t) = i(1+t)/(1-t)$, $t = (z-i)/(z+i)$ maps $D : |t| < 1$ onto H. Applying Schottky's theorem to $g(\phi(t))$ which omits 0 and 1 in D shows that there is a constant K such that

$$|g(t)| < \exp\left(\frac{K}{1 - |t|}\right)$$

If $z = re^{i\theta}$, $|t| < 1$, then

$$1 - |t|^2 = \frac{4r \sin\theta}{r^2 + 2r\sin\theta + 1}$$

whence

$$1 - |t| > \frac{2r\sin\theta}{r^2 + 2r\sin\theta + 1}$$

Thus

$$\log|g(re^{i\theta})| < \frac{K}{1 - |t|} < \frac{K(r+1)^2}{2r\sin\theta} < \frac{Kr}{\sin\theta}$$

if $r > 1$.

The next result is substantially contained in Edrei [4], but we give a proof for completeness.

LEMMA 5. Suppose g is entire and has only real zeros. Suppose further there are two values a, b such that $ab \neq 0$ and all the a-points of g are in $\operatorname{Im} z \geq 0$ while all the b-points are in $\operatorname{Im} z \leq 0$. Then g is at most of order one and exponential type.

Proof. By lemma 4 there is a constant K such that for all $r > 1$, $\theta \neq 0,\pi$

$$\log|g(re^{i\theta})| < Kr/|\sin\theta|$$

With the usual notations of Nevanlinna theory we have for $0 < \delta < \frac{\pi}{2}$

$$T(r,g) = \frac{1}{2\pi}\int_0^{2\pi} \log^+|g(re^{i\theta})|d\theta \tag{2}$$

$$\leq \frac{Kr}{\sin\delta} + \frac{1}{2\pi}\left[\int_{-\delta}^{\delta} + \int_{\pi-\delta}^{\pi+\delta} \log^+|g(re^{i\theta})|d\theta\right]$$

$$\leq \frac{Kr}{\sin\delta} + \frac{2\delta}{\pi}\log M(r,g)$$

Set $\phi = r/(\log T(r))$, $\delta = \{T(r)\}^{-\frac{1}{2}}$. Then for large r

$$\log M(r,g) \leq \frac{2r+\phi}{\delta} T(r+\phi) < \frac{3r}{\phi} T(r+\phi),$$

and (2) gives

$$T(r,g) < \frac{\pi}{2} Kr\{T(r)\}^{\frac{1}{2}} + \frac{6(\log T(r))}{\pi}\{T(r)\}^{-\frac{1}{2}}T(r+\phi) \tag{3}$$

A lemma of Borel [2, p. 18] states: For any increasing $V(r)$ which is continuous in $r > r_1$ and such that $V(r) \to \infty$ as $r \to \infty$ and for any $\eta > 0$ we have

$$V\left(r + \frac{r}{\log V(r)}\right) \leq (V(r))^{1+\eta}$$

outside a set of finite logarithmic measure.

Taking $V = T$ and $\frac{1}{2} > \eta > 0$ (3) shows that outside a set E of finite logarithmic measure

$$T^{\frac{1}{2}}(r,g) < \frac{\pi}{2} Kr + \frac{6}{\pi} (\log T(r,g))T^{\eta}(r,g)$$

so that

$$T^{\frac{1}{2}}(r)\{1 - o(1)\} < \frac{\pi}{2} Kr,$$

and for large $r \notin E$

$$T(r) < Ar^2, \quad A = (\pi K)^2$$

Since E has finite logarithmic measure, there exists a sequence $r_n \notin E$, $r_n \to \infty$ such that $1 < r_{n+1}/r_n < 2$. For $r_n \leq r \leq r_{n+1}$ we have

$$T(r) \leq T(r_{n+1}) \leq Ar_{n+1}^2 < 4Ar^2$$

Thus T has finite order ≤ 2.

But the growth of g in the angles $|\arg z \pm \frac{\pi}{2}| < \frac{\pi}{2} - \varepsilon$, for any $0 < \varepsilon < \frac{\pi}{4}$, is bounded by

$$|g(z)| < \exp\left(\frac{K|z|}{\sin \varepsilon}\right) \tag{4}$$

The Phragmèn-Lindelöf principle then shows that (4) holds for all θ as $z = re^{i\theta} \to \infty$ and g is of exponential type and order one.

LEMMA 6. Lehto [7]. If g(z) is entire, then

$$\limsup_{z\to\infty} |z|\rho(g(z)) = \infty$$

Hence by lemma 2 there is a sequence z_n such that $|z_n| \to \infty$ and Picard's theorem holds for g in any union of the discs

$$C_k = \{z : |z-z_k| < \varepsilon|z_k|\}$$

for each $\varepsilon > 0$.

LEMMA 7. Let f satisfy the assumptions of Theorem B. Then f is at most of order one and exponential type. If the lines ℓ_1, ℓ_2, ℓ_3 are all different, then at least two are parallel.

Proof. If the lines are different, the growth of f follows from lemma 3 with k = 1 while the case $\ell_1 = \ell_2$ follows from lemma 5 (with a = b) after a change of coordinates. The assumption that ℓ_1, ℓ_2, ℓ_3 are different and non-parallel leads to a contradiction with the second part of lemma 6 if we choose the constant $\varepsilon > 0$ sufficiently small.

LEMMA 8. Let g be entire of genus at most one. Suppose all zeros a_n of g be on Re z = r. Then

$$\overline{g(\overline{z} + r)} = g(-z + r)\exp(2Cz + iC') \tag{5}$$

and

$$\mathrm{Re}\{g'(z)/g(z)\} = C + \sum_n \frac{\mathrm{Re}\, z - r}{|z - a_n|^2} \tag{6}$$

for suitable real constants C and C'.

COROLLARY. If there is at least one zero, then $|g(z)| \to \infty$ in at least one of the cases $z = x \to \infty$ or $z = x \to -\infty$. If there are at least k zeros, then $|g(x)/x^{k-1}| \to \infty$ in at least one of these cases.

This lemma is given by Kobayashi [5]. The relation (6) follows immediately on calculation. If $r \neq 0$ or if $r = 0$, $g(0) \neq 0$, we may write

$$g(z) = e^{Kz} \prod_n \{(1 - \frac{z}{a_n})\exp(\frac{z}{a_n})\}$$

and

$$\frac{g'(z)}{g(z)} = K + \Sigma(\frac{1}{z - a_n} + \frac{1}{a_n})$$

whose real part gives (6) with $C = \operatorname{Re} K + \sum_n \operatorname{Re}(\frac{1}{a_n})$.

If $a_n = r + i\beta_n$, one has

$$\overline{g(\bar{z}+r)}/g(-z+r) = \exp\{(\bar{K} + K)z + (\bar{K} - K)\} \times \prod_n \left[(\frac{i\beta_n+r}{i\beta_n-r})\exp\{2z \operatorname{Re}(\frac{1}{a_n}) - 2\operatorname{Im}(\frac{1}{a_n})\} \right]$$

which has the form $\exp(2Cz + iC')$ where C' is real. Slight modifications in the formulae lead to the same results of $r = 0$ and $g(0) = 0$.

To prove the corollary, note that if $C > 0$ then $x > r$ implies $\operatorname{Re}\{g'(x)/g(x)\} < C$ and so $|g(x)| > |g(r)| \exp C(x-r)$.

If $C < 0$ then $x < r$ implies $\operatorname{Re}\{g'(x)/g(x)\} < C$ so that $\log|g(r)/g(x)| < C(r-x)$ and $|g(x)| > |g(r)|\exp C(x-r)$.

It only remains to prove the corollary if $C = 0$ and g has at least k zeros a_n. In this case we have for $x > r$ that

$$\operatorname{Re} g'(x)/g(x) \geq \sum_{n=1}^{k} \frac{x - r}{(x-r)^2 + \beta_n^2}$$

which implies

$$|g(x)/g(r)| \geq \prod_{n=1}^{k} |x - a_n|$$

and $\lim_{x\to\infty} (|g(x)|/x^{k-1}) = \infty$

LEMMA 9. Under the assumptions of lemma 8, assume further that g has finite exponential type. Then $\operatorname{Re}\{g'(x)/g(x)\}$ is bounded as real $x \to \pm\infty$.

Proof. We may take $r = 0$ and have to show that

$$S = \Sigma \frac{x}{x^2 + \beta_n^2}$$

is bounded for real x. Now $S = S_1 + S_2$ where (taking $x > 0$)

$$S_1 = \sum_{|\beta_n| \leq x} \frac{x}{x^2 + \beta_n^2} - \frac{n(0,x)}{x} = O(1) \text{ as } x \to \infty$$

since g has finite exponential type.

$$S_2 < 2x \int_x^\infty \frac{tn(0,t)dt}{(x^2 + t^2)^2} = x\, O \int_x^\infty t^{-2} dt = O(1)$$

as $x \to \infty$.

COROLLARY. For fixed real h

$$g(x + h)/g(x) = O(1) \qquad \text{as } x \to \pm\infty$$

Proof of Theorem B. From now on we assume that f satisfies the hypotheses of B, so by lemma 7 it has finite exponential type. If one of the values w_i, say w_1, is taken only finitely often, f has the form

$$f(z) = w_1 + P(z)\exp(Az + B)$$

where A, B are constants and P(z) is a polynomial of degree $p > 0$. If $p > 0$, the assumption that the roots of $f(z) = w_2$ lie on a line leads to a contradiction. For the roots can then be written $z_i = \alpha + \beta t_i$ where α, β are constant and real $t_i \to \infty$. The equation $f(z_i) = w_2$ leads to $\exp(t_i \mathrm{Re} A\beta) \sim k t_i^{-p}$ for some positive constant k, as $t_i \to +\infty$, which is impossible. Thus $p = 0$ and f has a form included in that asserted by the theorem.

Thus it can be assumed from now on that each ℓ_i contains an infinity of w_i-points of f.

LEMMA 10. If ℓ_1 and ℓ_2 are the lines $\mathrm{Re}\, z = h_1$ and $\mathrm{Re}\, z = h_2$ respectively, then for all z and for certain real constants C_1, C_2, C_1', C_2'

$$\overline{f(\bar{z}+h_j) - w_j} = \{f(-z+h_j) - w_j\}\exp(2C_j z + iC_j'), \; j = 1,2 \tag{7}$$

and if $k = h_2 - h_1$

$$f(z+2k) - w_2 = -\bar{w}_2 \exp\{-2C_2(2h_1-h_2) + 2C_2 z + iC_2'\} \qquad (8)$$
$$+ \{f(z)-w_1\}\exp\{2(C_2-C_1)z + 2C_2 k + i(C_2'-C_1')\}$$

Proof. (7) is lemma 8 applied to $f(z) - w_i$, which is of order one and hence of genus at most one. The relation (8) then follows by combining the cases $j = 1$ and 2 of (7).

LEMMA 11. If in lemma 10 $C_1 \neq C_2$, then f has an angular limit of w_1 or w_2 in either a left or right half-plane.

Proof. By lemma 8 we can assume $|f(x)| \to \infty$ either for real $x \to +\infty$ or for $x \to -\infty$. By changing z to -z, if necessary, we can assume that $|f(x)| \to \infty$ as $x \to +\infty$. By moving the origin we can assume $h_1 = 0$, $k = h_2$. This does not change the equality or otherwise of $C_i = \text{Re } f'(h_i)/\{f(h_i) - w_i\}$.

Suppose $C_2 > C_1$. Then (8) and the corollary to lemma 9 give as real $x \to +\infty$

$$\{1 + o(1)\}(f(x)-w_1)\exp\{2(C_2-C_1)x + 2C_2 k + i(C_2'-C_1')\} \qquad (9)$$
$$= -w_2 + \bar{w}_2 \exp\{2C_2 k + 2C_2 x + iC_2'\}$$

Both sides are unbounded as $x \to \infty$, so $C_2 > 0$ and hence

$$f(x) - w_1 \sim \bar{w}_2 \exp\{2C_1 x + iC_1'\} \qquad \text{as } x \to \infty \qquad (10)$$

Thus $C_2 > C_1 > 0$ and (7) with $j = 1$, $x > 0$ gives

$$f(-x) - w_1 = \exp\{-2C_1 x - iC_1'\}\{\overline{f(\bar{x})} - \bar{w}_1\} \qquad (11)$$
$$\sim w_2 \quad , \qquad x \to \infty$$

by (10).

If on the other hand $C_1 > C_2$ (8) yields as $x \to +\infty$

$$f(x + 2k) \sim -\bar{w}_2 \exp\{2C_2 k + 2C_2 x + iC_2'\}$$

since the left hand side must be unbounded.

Applying (7) for $j = 1$, we have for $x > 0$ in (11)

$$f(-x) - w_1 \sim \exp\{-2C_1 x - iC_1'\}(-)w_2 \exp\{2C_2 x - 2C_2 k + C_2'\}$$

and so $f(-x) - w_1 \to 0$ as $x \to +\infty$.

Suppose $f(x)$ has a constant limit λ as $x \to +\infty$. Since f omits the values w_1, w_2, and ∞ in the half-plane $\text{Re } z > \text{Max}(h_1,h_2)$, it is a normal function and has angular limit λ at ∞ in the half plane. The lemma is now established.

LEMMA 12. If the three lines ℓ_1, ℓ_2, and ℓ_3 are distinct, then by a rotation of the z-plane we may take at least two, say ℓ_1 and ℓ_2, to be parallel to the imaginary axis and assume that in the notation of lemma 11 $C_1 = C_2$.

Proof. If the ℓ_i are distinct, we can assume by lemma 1 that ℓ_1, ℓ_2 are parallel to the imaginary axis. If $C_1 \neq C_2$, then by lemma 11 f has angular limit which we may take to be w_1 in say a left half-plane and angular limit ∞ in a right half-plane (or conversely). Since there is an infinity of w_3-points on ℓ_3, this can happen only if ℓ_3 is parallel to ℓ_2.

But then the pair ℓ_2,ℓ_3 must have $C_2 = C_3$ by lemma 11.

LEMMA 13. If ℓ_1 and ℓ_2 are distinct and parallel to the imaginary axis with $C_1 = C_2 = C$, then theorem B holds.

Proof. By a translation of axes we may suppose ℓ_1 to be $\text{Re} z = 0$, ℓ_2 to be $\text{Re} z = k > 0$. Put $f(z) - w_1 = F(z)$ and $w_2 - w_1 = a$. Then lemma 10 shows that

$$F(z+2k) - a = F(z)e^{2Ck+i\delta} - \overline{w}_2 e^{2Ck+2Cz+iD} \tag{12}$$

where $\delta = C_2'-C_1'$ and $D = C_2'$ are real.

We dispose first of the special case when $C = 0$, $e^{i\delta} = 1$. In this case (12) becomes

$$F(z+2k) - F(z) = a - \overline{w}_2 e^{iD}$$

The right hand side is a constant which we call λ. Then $F(z) = \frac{\lambda z}{2k} + H(z)$, where $H(z)$ is periodic of real period 2k. Then F and hence f are $O(x)$ for real $z = x$. But $f(z) - w_1$ has an infinity of zeros on ℓ_1, so by lemma 8 $f(x)$ is not $O(x)$ for real x. Thus this case cannot occur.

Since we do not have both $C = 0$, $e^{i\delta} = 1$, we have $e^{2Ck} - 3^{i\delta} \neq 0$ and can find A such that $A(e^{2Ck}-e^{i\delta}) - \overline{w}_2 e^{iD} = 0$. With this choice of A, put $F(z) = Ae^{2Cz} + g(z)$ in (12) and we find

$$g(z+2k) - a = g(z)e^{2Ck+i\delta}$$

Choose μ so that

$$\mu - a = \mu e^{2Ck+i\delta}$$

which is possible since $e^{2Ck} \neq e^{-i\delta}$. Put

$$g(z) - \mu = p(z).\ \exp\left[\left(C + \frac{i\delta}{2k}\right)z\right]$$

which implies

$$p(z+2k) = p(z)$$

Since p is entire of order 1 and finite type, it has the form

$$p(z) = \sum_{m=-N}^{N} c_m \exp(m\pi iz/k), \qquad c_m \text{ constant}$$

and we have

$$f(z) = (w_1+\mu)+Ae^{2Cz} + e^{Cz} \sum_{m=-N}^{N} c_m \exp\{2m\pi i2 + i\delta z)/2k\} \tag{13}$$

If for any m both $c_m \neq 0$ and $m\pi+\delta \neq 0$ hold, then $|f(x_0 + iy)| \to \infty$ in at least one of the cases $y \to +\infty$ or $y \to -\infty$, say the case $y \to +\infty$. Then ℓ_1 and ℓ_2 contain no w_i-points for which $y > y_0$, say, and f omits w_1, w_2 in the plane except for the strip bounded by ℓ_1 and ℓ_2 in which $y \leq y_0$. From Schottky's theorem it follows that $|f(iy)|$ is o $\exp(|y|^{\frac{1}{2}})$ as $y \to \infty$. But this conflicts with the assumption that there is a term on the right hand side of (13) for which $|f(iy)| \to \infty$ as $y \to \infty$, for in this case $|f(iy)|$ must grow at least with order $\exp(qy)$ for some $q > 0$. Thus the last sum on the right hand side of (13) is empty, except perhaps for a term Be^{Cz}, B constant, and f has the form asserted in the theorem.

Theorem B is now established in the case when the lines ℓ_1, ℓ_2, and ℓ_3 are distinct, by lemmas 12 and 13. It remains to discuss the case when, say, $\ell_1 = \ell_2$. The general case will follow by simple transformations if we settle the case when $w_1 = 0$, $w_2 + 1$, and ℓ_1 is the real axis and by hypothesis $w_3 = c$ is not real. There are two alternatives to consider.

Case 1. Suppose f is real on the real axis. Then $\ell_3 \neq \ell_1$. Now ℓ_3 contains an infinity of c-points of f and the reflection ℓ_4 of ℓ_3 in ℓ_1 contains an infinity of $\bar{c}$-points ($\bar{c} \neq c$) which conflicts with the fact that f must have an angular limit ∞ in either the upper or lower half-plane (cf. lemma 8) unless ℓ_3 is parallel to ℓ_1. Then ℓ_2, ℓ_3, ℓ_4 are three distinct lines on which 0, c, and $\bar{c}$ are distributed so we have the case already dealt with and the theorem is proved.

Case II. f is not real for all points on the real axis. Edrei [4] by a simple distinction from the result of lemma 5 has shown that if f = 0, 1 only on the real axis, it has one of the forms

$$f(z) = \frac{\sin(\xi z-\eta)e^{i(\xi z-\eta)}}{\sin(\eta-\eta_1)}, \qquad \sin(\eta-\eta_1) \neq 0 \tag{14}$$

where ξ, η, and η_1 are real constants, or

$$f(z) = \frac{\sin p(\xi z+\eta)e^{i(p-1)(\xi z+\eta)}}{\sin(\xi z+\eta)} \tag{15}$$

where $p \neq 0, 1$ is an integer and ξ, η are real constants.

(14) is already of the form asserted by theorem B so we have only to consider f of the form (15).

If $\xi > 0$ then $f(iy) \to 1$ as $y \to \infty$ and $f(iy) \to \infty$ as $y \to -\infty$, while if $\xi < 0$ we have $f(iy) \to \infty$ as $y \to \infty$ and $f(iy) \to 1$ as $y \to -\infty$. Thus the line ℓ_3 must be parallel to ℓ_1. However $\ell_3 \neq \ell_1$, for if $\ell_1 = \ell_3$ then $\frac{1}{c}$ f takes 0,1 only on the real axis ℓ_1 which implies $\frac{1}{c}$ f(iy) $\to$ 1 either for $y \to \infty$ or $y \to -\infty$.

Thus f(iz) takes 0 on ℓ_1, the imaginary axis, and takes c on a line ℓ_4 parallel to ℓ_1 but distinct from it. The angular limits of f(iz) in left and right half-planes exist and are 1 and ∞. By lemma 11 $C_1 = C_4$ (in the notation of lemma 10 for f(iz)) and by lemma 13 f(iz) is quadratic in e^{Az} for some constant A. Theorem B is now established in all cases.

REFERENCES

1. I. N. Baker, Entire functions with linearly distributed values, *Math. Zeit, 86* (1964), 263-267.

2. E. Borel, Leçons sur les fonctions entières, 2ed. Paris 1921.

3. J. Clunie and W. K. Hayman, The spherical derivative of integral and meromorphic functions, *Comm. Math. Helv., 40* (1966), 117-148.

4. A. Edrei, Meromorphic functions with three radially distributed values, *Trans. Amer. Math. Soc., 78* (1955), 276-293.

5. T. Kobayashi, On a characteristic property of the exponential function, *Kōdai Math. Sem. Rep., 29* (1977), 130-156.

6. T. Kobayashi, Entire functions with three linearly distributed values, *Kōdai Math. J.*, *1* (1978), 133-158.

7. O. Lehto, The spherical derivative of a meromorphic function in the neighbourhood of an isolated singularity, *Comm. Math. Helv.*, *33* (1959), 196-205.

8. M. Ozawa, A characterization of the cosine function by the value distribution, *Kōdai Math. J.*, *1* (1978), 213-218.

ENTIRE FUNCTIONS WITH COMMON PREIMAGES

F. Gross

University of Maryland - Baltimore County
Baltimore, Maryland; and
Naval Research Laboratory
Washington, D. C.

Charles F. Osgood

Naval Research Laboratory
Washington, D. C.

1. INTRODUCTION

Given sets $S ,S_2,\ldots,S_n$ for some integers and meromorphic functions f and g which satisfy $f^{-1}(S_i) = g^{-1}(S_i)$ for $i = 1,2,\ldots,n$, multiplicities included, what can be said about f and g?

THEOREM A. Let f_1 and f_2 be two meromorphic functions, and let $a_j (j = 1,2, \ldots,5)$ be five distinct complex numbers such that the equations

$$f_i - a_j = 0$$

have the same roots not counting multiplicity for $i = 1,2$ for each fixed j. Then f_1 is identically equal to f_2.

Let g(z) be any meromorphic function and let

$$Eg(S) = \bigcup_{a \varepsilon S} \{z|g(z) - a = 0\}$$

where any z which is a zero of multiplicity m is included in Eg(S), m times.

The following theorem provided by the first author in [2] is an extension of Theorem A.

THEOREM B. Let $S_i (i = 1,2,3)$ be distinct finite sets of complex numbers such that no set is equal to the union of the other two, and let $T_1 (i=1,2,3)$ be any finite sets of complex numbers having the same number of elements as $S_i (i = 1,2,3)$, respectively. Let f(z) and g(z) be two meromorphic functions satisfying the conditions

$$E_f(S_i) = E_g(T_i) \quad (i = 1,2,3)$$

and

$$E_{\frac{1}{f}}(\{0\}) = E_{\frac{1}{g}}(\{0\})$$

then f(z) and g(z) are algebraically dependent.

The more we know about the sets S_i and T_i, the more we can say about the specific algebraic relationship between f and g.

For example, it is shown in [2] for entire functions that if f(z) and g(z) are nonconstant entire functions with a common Picard exceptional point a, and if $S(\neq \{a\})$ is any finite set of complex numbers such that $E_f(S) = E_g(S)$, then $f(z) = e^{\emptyset(z)} + a$ and g(z) is either of the form $ce^{\emptyset} + a$ with $c^n = 1$ for some integer n, or of the form $Ke^{-\emptyset} + a$, where K is a constant and $\emptyset(z)$ is entire.

It is worth observing that this last result shows in particular that there exists an infinite discrete set which characterizes an entire function almost completely. In fact if I is an integer and f is entire such that $E_f(I) = I$, then $f(z) = Az + B$, where $A = \pm 1$ and B is an integer. To see this simply observe that the hypotheses imply that $e^{2\pi i f(z)}$ is equal to 1 if and only if $f(z) \varepsilon I$.

Generally, the more sets S_i in the hypothesis, the more restricted are the corresponding functions which attain them simultaneously. On the other hand, for larger sets S_i it may be possible to find larger classes of functions which attain them simultaneously. For example, when n = 3, $S_1 = \{a\}$, $S_2 = \{b\}$ and $S_3 = \{c,d\}$ such that $S_i \cap S_j = \emptyset$ for i = j, all entire solutions f,g can be characterized. In fact, the first author proved the following theorem in [2]:

THEOREM C. Let f and g be nonconstant entire functions and let $S_1 = \{1\}$, $S_2 = \{-1\}$ and $S_3 = \{a,b\}$, where a and b are arbitrary constants such $S_i \cap S_j = \emptyset$ for $i \neq j$. Suppose that $f(z) \varepsilon S_i$ if $g(z) \varepsilon S_i$ $(i = 1,2,3)$ with the same multiplicities. Then f and g must satisfy one of the following relations.

$$f = g$$
$$fg = 1$$

or

$$(f-1)(g-1) = 4$$

Of course if in Theorem C the set S_3 is deleted from the hypotheses, then one expects to find a much larger family of solution pairs f and g.

If, however, reasonable assumptions are made about one of the functions, say f, then one can say correspondingly more about g.

Along these lines Rubel and Yang [3] proved:

THEOREM D. If f is entire with the same zeros and ones (counting multiplicity) as sinz, then f(z) = sinz.

2. MAIN RESULTS

Extensions of the above results can be obtained by means of recent results of Osgood and Yang [4] or Green [5].

We demonstrate the method using the results of [4]:

LEMMA (Osgood and Yang). Let P and Q be any two polynomials of the same degree. If

$$f = \frac{e^{Q}-1}{e^{P}-1}$$

is entire, then $Q = NP+2\pi ni$ for some integers N and n.

Before stating the Theorem, the proof of which will demonstrate our method, we return for a moment to Theorem A. If in this Theorem f_1 and f_2 are assumed to be entire, then the conclusion follows with four points a_1, a_2, a_3, and a_4. This is the best one can hope to do, since for 3 points the conclusion of Theorem A would be false. Simply observe for any entire function α and the points 1, -1, and 0, the function e^{α} and $e^{-\alpha}$ satisfy the hypotheses of Theorem A and are in fact distinct. One can also show that any two entire f and g which satisfy the hypotheses for 1, -1, and 0 must be of the form e^{α} and $e^{-\alpha}$.

Is it natural to ask what happens in the case when instead of considering the 3 distinct sets {-1}, {1}, and {0}, we consider instead $S_1 = \{-1,1\}$ and $S_2 = \{0\}$? Do we still get the same forms for f and g?

We prove that for f and g of finite order this question can be answered in the affirmative.

THEOREM. Let $S_1 = \{-1,1\}$, $S_2 = \{0\}$. If f and g are entire and of finite order such that $f \in S_i$, if and only if $g \in S_i$ $(i = 1,2)$ with exactly the same multiplicities, then either $f = \pm g$ or we may write f and g in the form

$$f = e^{Q} \text{ and } g = e^{-Q}$$

where Q is a polynomial.

Proof. From the hypotheses of the Theorem we have

$$(1) \quad f^2 - 1 = (g^2-1)e^{\alpha}$$

and

$$(2) \quad f = ge^{\beta}.$$

We consider three cases.

Case (1). β is a constant, say $e^{\beta} = c \neq 0$, then we have

$$c^2g^2 - 1 = (g^2-1)e^{\alpha}$$

or

$$g^2 = \frac{1-e^{\alpha}}{c^2-e^{\alpha}}$$

Since g is entire, we must have $c^2 = 1$ or e^{α} is a constant. Thus, in any case, $g = \pm f$ or g and f are constants.

Case (2). Suppose α is a constant, i.e., $e^{\alpha} = c \neq 0$, then we obtain

$$g^2 = \frac{1-c}{e^{2\beta}-c}$$

Since g is entire, we must have $e^{2\beta}$ a constant, so again either $f = \pm g$ or f and g are constants.

Case (3). Suppose that both α and β are non-constant. Then since f and g are of finite order, we have

$$g^2 = \frac{1-e^{\alpha}}{e^{2\beta}-e^{\alpha}} = \frac{e^{-\alpha}-1}{e^{2\beta-\alpha}-1} = \frac{e^{P_1}-1}{e^{P_2}-1},$$

where P_1 and P_2 are polynomials with $\deg P_2 \geq \deg P_1$.

Clearly, all the zeros of $e^{P_1}-1$ must be multiple, except for those for which also $e^{P_2}-1 = 0$. Since $e^{P_1}-1$ has only finitely many multiple zeros; namely, those for which $P_1 = 0$, it follows that $\deg P_2 = \deg P_1$. Otherwise, the numerator would have infinitely many simple zeros, which is impossible.

From the lemma we stated above, it follows that

$$g^2 = \frac{e^{nP_2+2\pi mi}-1}{e^{P_2}-1} = \frac{e^{nP_2}-1}{e^{P_2}-1}$$

Let $P_2 = P$, so that we get

$$g^2 = \frac{e^{nP}-1}{e^{P}-1}$$

If $n \neq \pm 1$, then the numerator would have a factor $e^P - \xi$, where ξ is some root of unity other than 1. All the infinitely many zeros of $e^P - \xi$ would again have to be multiple, which again is impossible. Thus, we must have $n = \pm 1$.

If $n = 1$, we again have $g = \pm 1$, a constant. Thus, we must have $n = -1$ and

$$g^2 = \frac{e^{-P}-1}{e^{P}-1} = \frac{e^{-P}(1-e^{P})}{e^{P}-1} = -e^{-P}$$

so that $g = \pm ie^{-P/2}$.

It readily follows that $f = \pm ie^{P/2}$.

This completes the proof of our Theorem.

It is worth noting the following consequence of our Theorem.

COROLLARY. With f, g, S_1, and S_2 as in the Theorem, if 0 is actually attained by f, then $f = \pm g$.

The question of whether the conditions in our Theorem that f and g be of finite order can be removed remains open.

A method similar to the above can be used to extend the results of our Theorem with the set $\{0\}$ replaced by the set $\{a\}$ where a is any complex number other than 1 or -1. The results of this more general case, though similar, are somewhat cumbersome to state, because of the many exceptions that occur. In any case all the solutions f and g can be characterized in terms of a.

In conclusion we remark that, as we indicated earlier, this method can be used and improved upon with the help of Green's result [5]. If the map $(e^{\phi_1}\ c^{\phi_2}\ e^{\phi_3})$ is of finite order, i.e., $N(f,o,r) \leq O(r^n)$ for some n, then f is a linear combination of e^{ϕ_1}, e^{ϕ_2}, e^{ϕ_3} whenever $f^2 = e^{2\phi_1} + e^{2\phi_2} + e^{2\phi_3}$.

Extensions of Green's result for more general rational forms would make it possible to obtain results for more general classes of sets $S_i (i = 1,2,...,n)$.

For a variety of related results on this topic see [6].

REFERENCES

1. R. Nevanlinna, *Le Theorems de Picard-Borel et la Theorie des Fonctions Meromorphes,* Gauthier-Villars, Paris, 1929.

2. F. Gross, On the distribution of values of meromorphic functions, *Trans. Am. Math. Soc.*, *131*, No. 1 (1968), 199.

3. L. A. Rubel and C. C. Yang, On zero-one sets for entire functions, *Mich. Math. J.*, *20* (1973), 289-296.

4. C. F. Osgood and C. C. Yang, On the quotient of two integral functions, *J. of Math. Anal. and Its Applications, 59, No. 2* (1976).

5. M. Green, On the functional equation $f^2 = e^{2\phi_1} + e^{2\phi_2} + e^{2\phi_3}$ and a new Picard theorem, *Trans. AMS, 195* (1974), 223-230.

6. F. Gross, *Factorization of Meromorphic Functions,* U. S. Govt. Printing Office Publication, Math. Res. Center, 1972, 165-182.

NEW METHODS IN FACTORIZATION OF MEROMORPHIC FUNCTIONS AND EXTENSIONS OF NEVANLINNA THEORY

F. Gross

University of Maryland - Baltimore County
Baltimore, Maryland; and
Naval Research Laboratory
Washington, D. C.

Charles F. Osgood

Naval Research Laboratory
Washington, D. C.

1. INTRODUCTION

The central problem of factorization can be stated intuitively as follows:

Given a meromorphic function F, in how many ways can it be expressed as a composition of meromorphic functions.

To make this more precise we introduce the following two definitions.

Definition 1. A meromorphic function $h(z) = f(g(z))$ is said to have $f(z)$ and $g(z)$ as left and right factors respectively, provided that f and g are meromorphic. The above representation of h is said to be a factorization of h.

Definition 2. Two factorizations $h = f_1 o g_1$ and $h = f_2 o g_2$ of h are said to be equivalent if and only if there exists a bilinear function L such that $f_2 = f_1 o L$ and $g_2 = L^{-1} o g_1$, where L^{-1} denotes the inverse of L.

The question of whether one can derive a general method for determining which factorizations are possible among transcendental functions is much more difficult than the analogous problem for polynomials. In fact even the functional equation

$$P(f) = Q(g) \tag{1}$$

where P and Q are centain specific classes of polynomials is very difficult to solve for f and g.

In this paper we shall discuss some new techniques for finding "large" meromorphic solutions f and g of (1) where the coefficients of the polynomials P and Q are "small" meromorphic functions. Generally, the methods used in solving factorization problems thus far have been based primarily on the classical theory of entire functions and on classical Nevanlinna Theory. The class of problems we shall discuss do not seem to lend themselves to the classical theory. Our efforts in solving this class of problems have led to some new and important extensions and generalizations of classical Nevanlinna Theory.

2. FERMAT EQUATIONS

Factorization problems are somewhat deceptive, in the sense that special classes such as the class described by equation (1) seem to be nothing more than a particular exercise. In reality these problems are quite general and very difficult.

If for example in (1) we take $P(W) = W^n$ and $Q(W) = 1 - W^n$, where n is any positive integer, then (1) becomes the classical Fermat equation:

$$f^n + g^n = 1 \tag{2}$$

The general solutions of (2) have been found for all n, but the methods used depend on uniformation theory as well as on some results from factorization.

In [1] and [2] the first author has made an extensive study of equation (2). It was shown in [1] that for n = 2 the most general meromorphic solutions f and g are given by

$$f(z) = \frac{2\beta(z)}{1+\beta(z)^2} \quad \text{and} \quad g(z) = \frac{1-\beta(z)^2}{1+\beta(z)^2}$$

where β is an arbitrary meromorphic function, while the most general entire solutions of (2) are given by

$$f = \sin\theta(z), \; g = \cos\theta(z)$$

where $\theta(z)$ is an arbitrary entire function. This last result was also proved by Jategaonkar [3] and Markushevich [4]. It was also noted in [1] that for $n > 2$, no nonconstant entire solutions exist, while for $n > 3$, no nonconstant meromorphic solutions exist. Furthermore, it was shown by the first author by means of certain factorization theorems [5], [6] and the uniformization theorem that the most general meromorphic solutions F and G of (2) for n = 3 are given by

$$F = f(h(z)) \text{ and } G = \eta g(h(z))$$

where $f = \frac{1}{2}(1 + \frac{P'}{\sqrt{3}})/P$, $g = \frac{1}{2}(1 - \frac{P'}{\sqrt{3}})/P$. P is a Weierstrass P-function satisfying the differential equation,

$$(P')^2 = 4P^3 - 1.$$

(i.e., $g_2 = 0$ and $g_3 = 1$), h(z) is any entire function and η is a cube root of unity. This result was also proved by Baker [7] independently.

If one considers variations of equation (2) by introducing certain classes of small coefficients, general solutions can still be obtained. In [8] for example the authors and Yang characterized all entire solutions of equations of the form

$$f^2 + qpq^2 = qq_1 \tag{3}$$

where p, q, and q_1 are relatively prime polynomials and q has no multiple zeros. The techniques of that paper did not yield any results on meromorphic solutions f and g of (3).

In [9] Heilbronn showed that the Fermat equation

$$f^3 + g^3 = z \tag{4}$$

has no entire solutions. His method also yields no results about meromorphic solutions.

One can, however, say a great deal about meromorphic solutions of (4) and more generally about solutions of

$$f^n + g^n = z \tag{5}$$

where n is any positive integer.

For example, it is obvious that for n > 3, (5) has no meromorphic solutions. Simply, replace z by z^n and divide by z^n, then (5) becomes

$$\left[\frac{f(z^n)}{z}\right]^n + \left[\frac{g(z^n)}{z}\right]^n = 1 \tag{6}$$

and we already know that for $n \geq 4$, no nonconstant solutions exist. Thus $f(z^n) = cz$. This implies that $f \equiv 0 \not\equiv c$ which of course yields no solutions to equation (5).

For n = 3, we can replace z by z^3 and obtain

$$\left(\frac{f(z^3)}{z}\right)^3 + \left(\frac{g(z^3)}{z}\right)^3 = 1$$

We may assume without any loss of generality that

$$\frac{f(z^3)}{z} \text{ has the form } \frac{1}{2}\left(1 + \frac{P'(h)}{\sqrt{3}}\right)/P(h)$$

and that

$$\frac{g(z^3)}{z} \text{ has the form } \frac{\eta}{2}\left(1 - \frac{P'(h)}{\sqrt{3}}\right)/P(h)$$

where, recall that η is a cube root of unity, P is a Weierstrass P-function, and h is entire.

Using these results, a careful analysis of the possible forms of f and g as well as the properties of the P-function lead to the most general solution of equation (5). Similar arguments can be used for the case when n = 2. More generally, arguing in some detail along the lines outlined above, one obtains

THEOREM 1. For $n \leq 3$, the equation

$$f(z)^n + g(z)^n = z^k \tag{7}$$

has nonconstant meromorphic solutions f(z) and g(z) for any integer k and all such solutions can be explicitly exhibited. For n > 3, no solutions exist unless k = n and f and g are linear.

The details of the proof of this theorem are presented in [10] by the authors.

One important consequence of Theorem 1 is that every entire function can be written as a sum of two cubes of meromorphic functions. For if h is any entire function, then replace z by h(z) in equation (4) and our assertion follows from Theorem 1.

3. NEW METHODS FOR FERMAT EQUATIONS WITH SMALL COEFFICIENTS

In order to get a handle on the problem of finding meromorphic solutions of equation (1), it is necessary to develop new methods or extensions and generalizations of the classical Nevanlinna theory. The authors have begun such an effort and have already succeeded in proving the following result:

THEOREM 2. Let

$$P(X,Y) = Q(X,Y), \tag{8}$$

where P and Q are each polynomials in X and Y having meromorphic coefficients, P is assumed to be homogeneous in X and Y of degree $K \geq 2$ with no repeated factors involving X or Y, $Q \not\equiv 0$ has degree $K_1 > K$ in X and Y, and P and Q have no common factors involving X or Y. Suppose that (8) has a solution $X = f$ and $Y = g$, where f and g are each meromorphic functions in the plane. Suppose further that each coefficient of (8) has a Nevanlinna characteristic function of size

$$o\Big(T(r,f) + T(r,g)\Big), \text{ then } K_1 > \left[\frac{K}{2} - 2\right]$$

where the square brackets denote the greatest integer's function.

A particular consequence of this result is that for $n \geq 5$, the equation

$$af^n + bg^n = 1 \tag{9}$$

has no meromorphic solutions satisfying the conditions

$$T(r,a) + T(r,b) = o\Big(T(r,f) + T(r,g)\Big)$$

Of course, solutions are known to exist for $n \leq 3$, at least for certain a and b. Two questions, however, remain open:

(a) Exactly for which small functions a and b do solutions exist when $n \leq 3$?

(b) Do there exist solutions when $n = 4$?

We shall prove in a subsequent paper that there are in fact interesting functions a and b for which solutions do exist and there are also pairs for which they do not exist. In fact, for appropriate choices of a, b, the solutions of equation (9) characterize some important classes of special functions.

We shall also show in a subsequent paper that for appropriate a and b analytic solutions of (9) exist for every positive integer. It seems, however, that these solutions are of bounded characteristic. One suspects that with suitable growth restrictions on solutions of f and g, solutions for $n > 3$ will not exist.

The result mentioned above regarding equation (8), though of considerable interest is not our primary objective. Our more important objective is in finding a more general method, a more global approach than has been available thus far to problems involving factorization and functional equations.

During the past year, we have been investigating this problem of finding a new approach for solving functional equations and factorizing certain classes of meromorphic functions which did not seem to yield to the classical theory. After much searching of the literature and agonizing over many problems, we suddenly realized that the clues we were searching for were apparent all the time, provided that we looked at it from the proper perspective.

What was discovered was that certain number theoretic techniques already discovered by the second author [11, 12, 13, 14] and used in dealing with the problem of establishing effective bounds upon the size of the polynomial or rational function points on certain curves defined on a field K(z), where generally K has characteristic zero, lead in a natural way to statements about large meromorphic points on curves defined over a field of small meromorphic functions. Upon careful scrutiny of the second author's work, one recognizes that this notion of "size" appearing in the rational function case was essentially the Nevanlinna characteristic function. With this observation made, the second author was able to derive a whole range of deficiency and/or ramification results for values which are not complex numbers but small functions. In particular, substantial generalizations of the three small function theorem [14] and its subsequent generalization by Chuang as well as a generalization of Milloux's Theorem [15] follow smoothly. The number theoretic technique of the second author depends heavily on certain especially constructed auxiliary differential equations. In [10] we develop this technique further and apply it to obtain our result regarding equation (8) mentioned earlier.

By means of Nevanlinna theory on more general Riemann Surfaces and certain facts on uniformization in algebraic geometry, we shall in subsequent papers construct appropriate auxiliary differential equations which will yield at least partial answers to the problems (a) and (b) posed above. Furthermore, we suspect that by generalizing the classical theory in an appropriate manner, we will succeed in looking correctly at the problem of solving at least some important classes of functional equations.

To better understand the directions that those generalizations might take, it is instructive to give a brief history of how the basic ideas which we hope to develop still further evolved.

Alan Baker suggested to the second author, some time ago, the problem of calculating effective bounds on the "diophantine approximation" of algebraic functions by rational functions (originally in the sense at the agreement of power series coefficients about $z = \infty$). The second author was soon able to provide very good bounds [11, 12]. Motivated by a paper of Kolchin [16], the second author introduced a new method to this area; he constructed an auxiliary differential equation which is satisfied by the algebraic function being approximated. Good bounds were obtained when the approximating rational functions could be shown not to satisfy the auxiliary differential equation. Generally, the better bounds to be established the more difficult the differential equation which had to be considered, i.e., the harder it was to make statements about its rational solutions. Thus, a range of bounds was obtained, depending upon the hypotheses.

Wolfgang Schmidt developed an interest in this area and has published a number of articles on it [17-21]. Schmidt has been particularly active in using and extending the second author's method to bound the size of rational or algebraic solutions of functional equations of type (8), where the small function coefficients are polynomials. Such bounds follow from good bounds in diophantine approximation, as they do for solutions of diophantine equations in classical number theory. In some ways these conculsions are stronger than those known for rational number solutions to the analogous diophantine equations.

In this context, Nevanlinna theory says that meromorphic functions behave very much like rational functions, especially with respect to differentiation. The Osgood-Schmidt arguments on approximation appear generally to map into arguments concerning the approximation of a small meromorphic or algebroid functions f (the analogues of rational functions with a large number of zeros and poles) in the sense of the size of the proximity function $m(r,\frac{1}{f-g})$ (the analogue of the order of vanishing of $f - g$ at $z = \infty$). The results bounding the size of solutions to certain functional equations of type (8) map over in a natural way [10]. Other arguments involving functional equations, that we have not yet thoroughly investigated, are also expected to map over.

Additionally, extensions of Nevanlinna theory to domains other than the plane exist and almost certainly the Osgood-Schmidt arguments will be applicable there. The requirements for the method to be applicable are fairly minimal: clearly the actual theorems which will result may differ with the geometry of the domain.

These investigations seem to bring together three basic areas of mathematics: number theory, complex variables, and - to an extent - geometry. The second author shows in [21] that under certain conditions he can nearly obtain the Nevanlinna constant of 2 for the sum of the deficiencies of f "at" small functions g (he obtains $2 + \varepsilon$ for each $\varepsilon > 0$). The proof depends upon utilizing an argument which was developed in [13] in hopes of establishing a Thue-Siegel-Roth theorem for algebraic functions.

The authors believe that the initial ideas developed by the second author and Schmidt and their further development by the authors of this paper [10] may well be the tip of an iceberg. The relationship between Number Theory, Complex Analysis, and Geometry referred to above if successfully unravelled will lead to a much clearer picture of what is generally happening and generalizations of the type discussed above will almost certainly follow.

REFERENCES

1. F. Gross, On the equations $f^n+g^n = 1$, *Bull. Amer. Math. Soc., 72,* No. 1, Part I (1966), 86-88.

2. F. Gross, On the equation $f^n+g^n = 1$, II, *Bull. Amer. Math. Soc., 74,* No. 4 (1968), 647-648.

3. A. V. Jategaonkar, Elementary proof of a theorem of P. Montel, *J. London Math. Soc., 40* (1965), 166-170.

4. A. I. Markushevich, *Entire Functions*, Chapter 5, Amer. Elsevier Pub. Co., NY, 1966.

5. F. Gross, On factorization of elliptic functions, *Canadian J. Math., 20* (1968), 486-494.

6. F. Gross, On factorization of meromorphic functions, *Trans. of the Amer. Math. Soc., 131* (1968), 215-221.

7. I. N. Baker, On a class of meromorphic functions, *Proc. Amer. Math. Soc., 17* (1966), 819-822.

8. F. Gross, C. F. Osgood, and C. C. Yang, On entire solutions of a functional equation in the theory of fluids, *J. Math. Phys., Vol. 16, X,* (1975).

9. W. K. Hayman, *Research Problems in Functions Theory,* Univ. of London, The Athlone Press, 1967, p. 16.

10. F. Gross and C. F. Osgood, On the functional equation $f^n+g^n = h^n$ and a new approach to a certain class of more general functional equations, *Indian Math. J.,* to appear.

11. C. F. Osgood, An effectively lower bound on the "Diophantine Approximation" of algebraic functions by rational functions, *Mathematika, 20* (1973), 4-15.

12. C. F. Osgood, On the effective "Diophantine Approximation" of algebraic functions over fields of arbitrary characteristic and applications to differential equations, *Proc. Kanink Nederl. Akad. Van Wetens,* Amsterdam, Series A, 78, No. 2 (1975), 105-119.

13. C. F. Osgood, Concerning possible "Thue-Siegel-Roth Theorem" for algebraic differential equations, *Number Theory and Algebra*, edited by Zassenhaus, Academic Press, 1977, pp. 223-224.

14. C. F. Osgood, An effective lower bound on the "Diophantine Approximation" of algebraic functions by rational functions II, *Contributions to Algebra*, Academic Press, 1977.

15. W. K. Hayman, *Meromorphic Functions*, Oxford Mathematical Monographs, Oxford at the Clarendon Press, 1964.

16. E. R. Kolchin, Rational approximation to solutions of algebraic differential equations, *Proc. Amer. Math. Soc., 10* (1959), 238-244.

17. W. M. Schmidt, Rational approximation to solutions of linear differential equations with algebraic coefficients, *Proc. Amer. Math. Soc., 53* (1975) 285-289,

18. W. M. Schmidt, On Osgood's effective Thue Theorem for algebraic functions, *Commun. on Pure and Applied Math., 29* (1976), 759-773.

19. W. M. Schmidt, "Diophantine Approximation" in power series fields, *Acta Arith., 32* (1977), 275-296.

20. W. M. Schmidt, Thue's Equation over function fields, *Aust. Math. Soc.*, Series A, No. 25 (1978), 385-422.

21. W. M. Schmidt, Contributions to "Diophantine Approximation" in fields of series, *Montsh. Math., 87* (1979), 145-165.

22. C. F. Osgood, A number theoretic - differential equations approach to generalizing Nevalinna Theory, *Indian J. of Math.*, to appear.

SOME PROPERTIES OF REAL ENTIRE FUNCTIONS

Tadashi Kobayashi

Chiba Keiai University
Chiba, Japan

1. INTRODUCTION

Let f(z) be a real entire function with only real zeros. Recall that a real entire function is one which assumes real values on the real axis. By E(f), we denote the set of real numbers t for which all the roots of the equations f(z) = t are real only. On the set E(f), the following facts have been proved by Edrei [2]:

(A) *If the set* E(f) *contains two points, then* f(z) *has at most order one and mean type.*

(B) E(f) *is a closed interval in the real field.*

(C) *If* E(f) *is unbounded, then* f(z) *reduces to a polynomial of degree at most two.*

For instance, let P(z) be a real polynomial of degree at least three, all of whose zeros are real and simple. Then by an elementary analysis, we can easily conclude that

$$E(P) = [\alpha, \beta]$$
$$\alpha = \max \{ P(x) : P'(x) = 0, P(x) < 0 \}$$
$$\beta = \min \{ P(x) : P'(x) = 0, P(x) > 0 \}$$

In this paper we investigate the possibility of proving analogous facts for the sets E(f) of real entire functions with only real zeros.

2. PRELIMINARY

Let f(z) be a real entire function having only real zeros. Assume that the order of f(z) is greater than one. Then by the above (A), the set E(f) consists of the single point 0. Hence we may confine ourselves to the class of real entire functions of genus at most one which have only real zeros. Hereafter we denote this class by A.

Let f(z) be a function of the class A. Then we have

$$\operatorname{Im} \frac{f'(z)}{f(z)} = -\Sigma \frac{\operatorname{Im} z}{|z - a_n|^2} \tag{2.1}$$

$$\frac{d}{dz} \frac{f'(z)}{f(z)} = -\Sigma \frac{1}{(z - a_n)^2} \tag{2.2}$$

for values of z with $z \neq a_n$, where the a_n are the zeros of f(z). By these (2.1) and (2.2), we have easily the following lemmas.

LEMMA 1. *Let* s *be a zero of* f'(z). *Then* s *must be real. Furthermore, if* f"(s) = 0, *then* f(s) = 0.

LEMMA 2. *Let* a *and* b *be successive zeros of* f'(z) *with* a < b *and* $f(a)f(b) \neq 0$. *Then there exists exactly one zero of* f(z) *in the open interval* (a,b).

Assume that $E(f) = [\alpha,\beta]$ for some function f(z) of the class A. Then it is clear by the definition that $E(-f) = [-\beta,-\alpha]$, E(f(-z)) = E(f), and E(f(z+c)) = E(f) for any real number c. Hence it is sufficient to consider the following three subclasses of A.

$A_1 = \{f(z) \in A$: f(z) has infinitely many positive and negative zeros, and $f(0) > 0\}$,

$A_2 = \{f(z) \in A$: f(z) has infinitely many positive zeros and $f(x) > 0$ for $x \leqq 0$,

$A_3 = \{f(z) \in A$: f(z) has finitely many zeros$\}$.

Firstly, let f(z) be a function of the class A_1. By $\{a_j\}$, we denote the zeros of f(z) and enumerate them such as

$$a_j \leqq a_{j+1}, \quad a_{-1} < 0 < a_0 \tag{2.3}$$

Here each zero appears as often as indicated by its multiplicity. Assume that $a_j < a_{j+1}$ for some integer j. Then by the above Lemma 2, there is exactly one zero of f'(z) in the open interval (a_j, a_{j+1}). We denote it by b_j. If

$$a_{j-1} < a_m = a_{j+1} = \cdots = a_{j+m-1} < a_{j+m} \qquad (m \geqq 2) \tag{2.4}$$

then we set

$$b_j = b_{j+1} = \cdots = b_{j+m-2} = a_j$$

Evidently $f'(b_j) = 0$ for each integer j, and the set $\{b_j\}$ surely coincides with the set of all the zeros of f'(z). By (2.3) and $f(0) > 0$, it follows that $a_{-1} < b_{-1} < a_0$ and $f(b_{-1}) > 0$. Further

$$a_{j-1} < b_{j-1} < a_j = b_j < b_{j+m-1} < a_{j+m}$$

in the case (2.4). Hence $f(b_{j-1})f(b_{j+m-1})$ is positive or negative according to whether m is even or odd. Therefore we obtain

$$(-1)^{j+1} f(b_j) \geqq 0$$

for each integer j. Here let us set

$$\begin{aligned} \alpha &= \sup f(b_{2j}) \leqq 0 \\ \beta &= \inf f(b_{2j+1}) \geqq 0 \end{aligned} \tag{2.5}$$

where j ranges over all integers.

Next we consider the class A_2. Let f(z) be a function of A_2, and let

$$0 < a_0 \leqq a_1 \leqq a_2 \leqq a_3 \leqq \cdots$$

be the sequence of its zeros, each zero appearing as often as its multiplicity indicates. As before, by $\{b_j : j=0,1,\ldots\}$, we denote the sequence of the zeros of f'(z) which lie in the semi-infinite interval $x \geqq a_0$. Then by means of Lemma 2, we can enumerate this sequence such as $a_j \leqq b_j \leqq a_{j+1}$ for every non-negative integer j. Further, by the nature of b_j, we also have

$$(-1)^{j+1} f(b_j) \geqq 0$$

for $j \geqq 0$. Hence we can set

$$\begin{aligned} \alpha' &= \sup f(b_{2j}) \leqq 0 \\ \beta' &= \inf f(b_{2j+1}) \geqq 0 \end{aligned} \tag{2.6}$$

where j ranges over all integers with $j \geqq 0$.

Now we consider the class A_3. Let f(z) be a transcendental real entire function of A_3. Then

$$f(z) = P(z)e^{az}$$

where P(z) is a real polynomial with only real zeros and a is a real constant with $a \neq 0$. Hence for each real number t, the equation f(z) = t has at most finitely many real roots. Therefore it is clear that E(f) = {0}. Hereby we may ignore this class A_3.

3. STATEMENT OF RESULTS

Our task is to prove the following facts.

THEOREM 1. *Let* f(z) *be a real entire function of the class* A_1. *Then the set* E(f) *coincides with the closed interval* $[\alpha\ \beta]$, *where* α *and* β *are the real numbers defined by* (2.5).

THEOREM 2. *Let* f(z) *be a real entire function of the class* A_2. *Then either*

$$\lim_{x\to-\infty} f(x) = \infty \tag{3.1}$$

or

$$\lim_{x\to-\infty} f(x) = 0 \tag{3.2}$$

(a) *In the case where* (3.1) *holds, the set* E(f) *coincides with the closed interval* $[\alpha',\beta']$, *where* α' *and* β' *are the real numbers defined by* (2.6).

(b) *If* (3.2) *holds, then the set* E(f) *consists of the single point* 0.

The next three theorems are immediate consequences of the above Theorems 1 and 2.

THEOREM 3. *Let* f(z) *be a real entire function of the class* A *for which the set* E(f) *has the point* 0 *as its interior point. Let* P(z) *be a real polynomial with only real zeros such that all the zeros of the function* g(z) = P(z)f(z) *are real and simple. Then the set* E(g) *is a closed interval which does not reduce to a point.*

THEOREM 4. *Let* f(z) *be a transcendental real entire function belonging to the class* A. *Assume that there is a positive number* c *such that*

$$(f(x))^2 + (f'(x))^2 \geqq c^2$$

for real values of x. *Then the sets* E(f) *and* E(f') *both contain the closed interval* [-c,c].

Theorem 5. *Let* f(z) *be a transcendental real entire function of the class* A, *whose zeros are all simple. Assume that the set* E(f) *contains more than two points and that there is a positive number* M *such that* $|a - b| \leqq M$ *for any successive zeros* a *and* b *of* f(z). *Then for each natural number* n, *the set* $E(f^{(n)})$ *has the point* 0 *as its interior point. Hence* $E(f^{(n)})$ *also contains more than two points.*

Furthermore, Theorems 1 and 2 suggest to us how to construct real entire functions f(z) for which the sets E(f) contain more than two points. In the final section, we shall present various kinds of such functions along this line.

4. PROOF OF THEOREM 1.

Let f(z) be a real entire function of the class A_1. As before, let $\{a_j\}$ and $\{b_j\}$ be the zeros of f(z) and f'(z), respectively. Then by what was mentioned in Section 2, we can enumerate them such as

$$a_j \leqq b_j \leqq a_{j+1}, \qquad a_{-1} < 0 < a_0 \tag{4.1}$$

for each integer j.

Assume that $a_j < a_{j+1}$ for some integer j. Then $a_j < b_j < a_{j+1}$ and $f(b_j)$ is positive or negative according to whether j is odd or even. Further, $f''(b_j) \neq 0$ by Lemma 1. Hence by the reality of f(z), we can define the curve $C_j : z = z_j(t)$ $(0 \leqq t < r_j \leqq +\infty)$ which lies in the upper half plane H except for its initial point $z_j(0) - b_j$ and satisfies

$$\lim_{t \to r_j} z_j(t) = \infty$$

$$f(z_j(t)) = f(b_j) + (-1)^{j+1} t \tag{4.2}$$

for $0 \leqq t < r_j$, that is, the curve C_j reaches to the infinity and along which f(z) takes real values and converges to the value $f(b_j) + (-1)^{j+1} r_j$.

Next let us consider the case where

$$a_{j-1} < a_j = a_{j+1} = \dots = a_{j+m-1} < a_{j+m}$$

for some integers j and m with $m \geqq 2$. Then from the construction of the sequence $\{b_j\}$, it follows that

$$a_{j-1} < b_{j-1} < a_j = b_n < b_{j+m-1} < a_{j+m}$$

for $n=j,\ldots,j+m-2$. Since $f(z)$ has a zero of order m at the point a_j and $f'(z)$ has no zeros in the upper half plane H, we can also set the curves $C_n : z = z_n(t)$ $(0 \leqq t < r_n \leqq +\infty)$ for $n=j,\ldots,j+m-2$ such that

$$z_n(0) = b_n = a_j$$

$$\mathrm{Im}\, z_n(t) > 0 \qquad (0 < t < r_n)$$

$$\lim_{t\to r_n} z_n(t) = \infty$$

$$\lim_{t\to+0} \arg(z_n(t) - z_n(0)) = \frac{m-1-n+j}{m}\pi \tag{4.3}$$

and

$$f(z_n(t)) = f(b_n) + (-1)^{n+1} t = (-1)^{n+1} t$$

for $0 \leqq t < r_n$. Consequently, for each integer j we can define the curve $C_j : z = z_j(t)$ $(0 \leqq t < r_j \leqq +\ )$ which lies in H except for its initial point b_j, ends at the infinity, and satisfies (4.2). Since $f'(z) \neq 0$ in H, these C_j are simple and analytic curves, and they are mutually disjoint.

By D_j, we denote the simply connected subregion of the upper half plane H which is surrounded by C_j, C_{j+1}, and the segment b_jb_{j+1} of the real axis. Then from the construction of the curves C_j, in particular from (4.3), all the regions D_j are mutually disjoint.

4.1 Let us recall (2.1). Then

$$\mathrm{Im}\,\frac{f'(z_j(t))}{f(z_j(t))} < 0$$

for $0 < t < r_j$. On the other hand, it follows from (4.2) that

$$(-1)^{j+1} f(z_j(t)) > (-1)^{j+1} f(b_j) \geqq 0 \tag{4.4}$$

and

$$f'(z_j(t))z'_j(t) = (-1)^{j+1}$$

for $0 < t < r_j$. Hence $\mathrm{Im}\, z'_j(t)$ is always positive, so that $\mathrm{Im}\, z_j(t)$ is strictly increasing for $0 \leqq t < r_j$.

Now let us assume that $y_n(t) = \text{Im } z_n(t)$ $(0 \leqq t < r_n)$ is bounded above for some integer n. Then $x_n(t) = \text{Re } z_n(t)$ converges to $+\infty$ or $-\infty$ as t tends to r_n, because the curve C_n converges to the point at infinity. Hereafter we assume that $x_n(t) \to +\infty$ as $t \to r_n$. (In the case where $x_n(t) \to -\infty$, the argument is quite similar to that below.) Since the curves C_j are mutually disjoint and the functions $y_j(t) = \text{Im } z_j(t)$ are strictly increasing, we thus have

$$v_j = \lim_{t \to r_j} y_j(t) \leqq v_n = \lim_{t \to r_n} y_n(t)$$

$$\lim_{t \to r_j} x_j(t) = +\infty$$

for every integer j with $j \geqq n$.

Here we further assume that $v_n > v_{n+1}$. Then the curve C_n meets the line $\text{Re } z = v_{n+1}$ at only one point p, and the half line $I = \{x + iv_{n+1} : x > \text{Re } p\}$ is contained entirely in the region D_n. Let us recall (2.1) again. Then it is possible to find a point z* of I such that the value f(z*) is real and positive. By E(w,f(z*)), we denote the element of the inverse function of f(z) with center f(z*) and satisfying E(f(z*),f(z*)) = z*. We now continue analytically this element E(w,f(z*)) along the real axis from the point f(z*) toward 0. Then we have an analytic continuation up to some point u $(0 \leqq u < f(z^*))$, with the possible exception of this end point. Thus from this continuation, we can define the simple curve $C : z = z(t)$ $(0 \leqq t < f(z^*)-u)$ such that $z(0) = z^*$ and

$$f(z(t)) = f(z^*) - t \tag{4.5}$$

for $0 \leqq t < f(z^*)-u$. Since f(z) assumes only real values on C, from the construction of the curves C_j, this curve C does not meet any curves C_j and the real axis. Hence we may assume that the above continuation defines a transcendental singularity at the point u, so that z(t) tends to infinity as t does to $r = f(z^*)-u$. Of course, the curve C is an asymptotic path and is contained entirely in the region D_n. On the other hand from (2.1) and (4.5), the imaginary part of z(t) decreases strictly as t varies from 0 to r. It thus follows that

$$\lim_{t \to r} \text{Im } z(t) < \text{Im } z(0) = v_{n+1}$$

Hence C lies in a bounded set, so that C does not approach the point at infinity. This is a contradiction. Consequently we have $v_n = v_{n+1}$. Therefore by induction, $v_j = v_n$ for $j \geqq n$. Furthermore, by the exact same way as above, we can also prove that

$$\begin{aligned} &\lim_{t \to r_j} y_j(t) = v_n \\ &\lim_{t \to r_j} x_j(t) = +\infty \end{aligned} \tag{4.6}$$

for $j \leqq n$. Hence all the curves C_j and all the regions D_j are contained in the strip $0 \leqq \text{Im } z \leqq v_n$.

Since $z_j(0) = b_j < b_0$ for $j \leqq -1$, every curve C_j ($j \leqq -1$) intersects the line Re $z = b_0$ by (4.6). Hence it is possible to take a point z_j of C_j such that Re $z_j = b_0$ and Im $z_j <$ Im z_{j-1} for $j \leqq -1$. From (4.6), it is clear that the sequence $\{z_j : j=-1, -2, \ldots\}$ converges to some point s in the upper half plane H. However, by (4.4) this limit point s must be a zero point of f(z). This is a contradiction.

Accordingly, we have proved the following.

LEMMA 3. *Let* $C_j : z = z_j(t)$ $(0 \leqq t < r_j \leqq +\infty)$ *be the curves defined above. Then for each integer* j, *the imaginary part of* $z_j(t)$ *is strictly increasing for* $0 \leqq t < r_j$ *and converges to infinity as* t *tends to* r_j.

4.2 From this Lemma 3, for an arbitrary positive number y^*, every curve C_n meets the line Im $z = y^*$ exactly once. By x_n+iy^*, we denote the intersection of C_n and I_m $z = y^*$. Then from $b_n \leqq b_{n+1}$ and from the fact that all the curves C_n are mutually disjoint, it follows that $x_n < x_{n+1}$ for each integer n. Further, the intersection of the region D_n and the line Im $z = y^*$ is the open segment between the points x_n+iy^* and $x_{n+1}+iy^*$. Assume now that the sequence $\{x_n\}$ is bounded above. Then x_n+iy^* clearly converges to some finite point x^*+iy^* as $n \to +\infty$. Hence we have $f(x^*+iy^*) = 0$ by (4.4). Since f(z) has no zeros in the upper half plane H, this is impossible. Hereby $x_n \to +\infty$ as $n \to +\infty$. Similarly, $x_n \to -\infty$ as $n \to -\infty$. By this fact, an arbitrary point of H must belong to some region D_n or some curve C_n.

In order to complete the proof of Theorem 1, it is sufficient to show the next lemma.

LEMMA 4. *Let* D_j *be the simply connected regions defined above. Then* f(z) *effects a one-to-one conformal mapping of the region* D_j *onto the upper half plane or the lower half plane according to whether* j *is even or odd.*

Proof. Assume that there exists a point z, in some region D_n at which f(z) takes a real value. Then $f(z_*) \neq 0$ by the assumption. Here we further assume that the value $f(z_*)$ is positive. (For the case where $f(z_*)$ is negative, the argument is quite similar to that below.) Then by the same reason as before, we can set the simple analytic curve $C^* : z = z^*(t)$ $(0 \leqq t < r^* \leqq f(z_*))$ which lies in the region D_n and satisfies $z^*(0) = z_*$,

$$\lim_{t \to r^*} z^*(t) = \infty \tag{4.7}$$

and

$$f(z^*(t)) = f(z_*) - t \tag{4.8}$$

for $0 \leqq t < r^*$. In other words, this curve C^* is an asymptotic path starting from the point z_*, and along which f(z) takes real positive values only and converges to $f(z_*)-r^*$. On the other hand from (2.1) and (4.8), the imaginary part of $z_*(t)$ is also decreasing for $0 \leqq t < r^*$, so that the curve C^* must be contained in the intersection of D_n and the strip $0 \leqq \mathrm{Im}\, z \leqq \mathrm{Im}\, z_*$. This is clearly absurd by (4.7). Consequently f(z) does not take any real values in each region D_j. Hence by (4.4), the inage $f(D_j)$ is contained in the upper half plane or the lower half plane according to whether j is even or odd. It therefore follows from the classical Lindelöf's theorem that f(z) approaches infinity as z tends to infinity along each curve C_j. Thus

$$(-1)^{j+1} f(b_j) + r_j = +\infty$$

so that the quantities r_j are all $+\infty$. Furthermore, since f'(z) has no zeros in the upper half plane H, the monodromy theorem yields that f(z) is univalent in each region D_j and the image $f(D_j)$ coincides with the upper half plane or the lower half plane according to whether j is even or odd. This is the desired result.

4.3 We shall now complete the proof of Theorem 1. Let t be an arbitrary real number with $t < \alpha$, where α is the quantity defined by (2.5). Then there is some even integer n such as $t < f(b_n) \leqq \alpha$. Hence from (4.2), the equation $f(z) = t$ has a root on the curve C_n, so that this equation has non real roots. Similarly for each real number t with $t < \alpha$, the equation $f(z) = t$ has also nonreal roots. Therefore, the set E(f) is a subset of the closed interval $[\alpha, \beta]$. Conversely, let s be a point of the upper half plane H such that the value f(s) is real. Then by virtue of the above Lemma 4, this point s must be a point of some curve C_n. Hence by (4.2), $f(s) \leqq f(b_n)$ or $f(s) \geqq$

$f(b_n)$ or $f(s) \geqq f(b_n)$ according to whether n is even or odd. Thus either $f(s) \leqq \alpha$ or $f(s) \geqq \beta$. Since f(z) satisfies

$$f(z) = \overline{f(\overline{z})}$$

it therefore follows that the equation $f(z) = t$ has only real roots for each real number t with $\alpha < t < \beta$. Hereby the set E(f) contains the interval (α,β). Consequently the set E(f) coincides with the closed interval between α and β. This completes the proof.

4.4 As an immediate consequence of this Theorem 1, we have the next fact.

COROLLARY. *Let* g(z) *be a real entire function of the class* A_1, *and let*

$$\cdots \leqq a_{j-1} \leqq a_j \leqq a_{j+1} \leqq \cdots$$

be the sequence of its zeros. Assume that all the zeros are simple, that is, $a_j < a_{j+1}$ *for every integer* j. *Then the following two statements are equivalent:*

(i) *The set* E(g) *has the point* 0 *as its interior point.*

(ii) *It is possible to take a point* c_j *in each interval* (a_j, a_{j+1}) *satisfying*

$$\inf_j |g(c_j)| > 0$$

where j *runs through all the integers.*

5. PROOF OF THEOREM 2

Let f(z) be a real entire function of the class A_2, and let

$$0 < a_0 \leqq a_1 \leqq a_2 \leqq a_3 \leqq \cdots$$

be the sequence of its zeros. Here each zero appears as often as indicated by its multiplicity. As before, by $\{b_j : j = 0, 1, \ldots\}$, we denote the sequence of the zeros of f'(z) which lie in the semi-infinite interval $x \geqq a_0$. Then we can enumerate this sequence such that $a_j \leqq b_j \leqq a_{j+1}$ for each nonnegative integer j. Since $f(0) > 0$ and $a_0 > 0$, it follows that $(-1)^{j+1}f(b_j) \geqq 0$ for $j \geqq 0$.

5.1 Firstly we consider the case where the genus of the sequence $\{a_n\}$ is one, that is,

$$\sum_{n\geqq 0} a_n^{-1} = +\infty \tag{5.1}$$

Then we have

$$f(z) = \exp(Az+B) \prod_{n\geqq 0} \left(1 - \frac{z}{a_n}\right)\exp\left(\frac{z}{a_n}\right)$$

with real constants A and B. Hence

$$\log f(x) = Ax + B - x^2 \int_0^{+\infty} \frac{n(t)}{t^2(t-x)}\,dt$$

$$\leqq Ax + B + \frac{x}{2} \int_0^{-x} \frac{n(t)}{t^2}\,dt$$

for negative real values of x, where n(t) is the counting function of the sequence $\{a_n\}$. Thus from (5.1), f(x) converges to 0 as x tends to infinity along the negative real axis.

Next we assume that the sequence $\{a_n\}$ satisfies

$$\sum_{n\geqq 0} a_n^{-1} < +\infty \tag{5.2}$$

Then f(z) can be written such as

$$f(z) = \exp(Az+B) \prod_{n\geqq 0} \left(1 - \frac{z}{a_n}\right)$$

with real constants A and B. Hence we have

$$\log f(z) = Az + B + z \int_0^{-\infty} \frac{n(t)}{t(z-t)}\,dt$$

except for real positive z. From this representation, it follows that

$$\lim_{x\to-\infty} f(x) = 0$$

or

$$\lim_{x\to-\infty} f(x) = +\infty$$

according to whether A > 0 or $A \leqq 0$. Here we further remark that

$$\log|f(x+iy)| \geqq Ax + B + \int_0^{-x} \frac{n(t)}{2t}\,dt$$

for real x and y with x < 0. Thus if $A \leqq 0$, then

$$\lim_{x\to-\infty} |f(x+iy)| = \infty \tag{5.3}$$

uniformly for real values of y.

5.2 Assume now that the function $f(z)$ satisfies

$$\lim_{x\to-\infty} f(x) = 0 \tag{5.4}$$

Then by what is mentioned just above, the genus of $f(z)$ must be one. Assume further that the set $E(f)$ contains two points a and b other than 0. Then by the definition, all the roots of the equations $f(z) = a$ and $f(z) = b$ are distributed only on the real axis. Moreover, from (5.4) it is possible to find a real number c such that the two equations $f(z) = a$ and $f(z) = b$ have no roots on the semi-infinite interval $x \leqq c$. Here let us consider the function defined by $g(z) = f(z^2+c)$. Then $g(z)$ is also real entire and all the roots of $g(z) = a$ and $g(z) = b$ must be real. Hence the set $E(g)$ also contains more than two points. Thus the proposition (A) implies $T(r,g) = O(r)$, where $T(r,g)$ denotes the characteristic function of $g(z)$. On the other hand, it is clear that

$$T(r,g) \sim T(r^2,f) \qquad (r \to +\infty)$$

Hereby $T(r,f) = O(r^{\frac{1}{2}})$, so that the genus of $f(z)$ must be zero. This is a contradiction. Consequently, if $f(z)$ satisfies (5.4), then the set $E(f)$ consists of the single point 0.

5.3 We shall now make a detailed study of the case where the function $f(z)$ satisfies

$$\lim_{x\to-\infty} f(x) = +\infty , \tag{5.5}$$

In this case $f(z)$ also satisfies (5.3). From Lemma 2, $f'(z)$ has at most one zero in the semi-infinite interval $x < a_0$. For a moment we assume that $f'(z)$ has a zero there. Then it must be simple by Lemma 1. Hence $f(x)$ is bounded for $x < a_0$, since $f(x)$ is positive and has the maximum there. This is clearly absurd by (5.5). Therefore all the zeros of $f'(z)$ are distributed only on the half line $x \geqq a_0$, so that $f'(z)$ vanishes only at the points b_j $(j \geqq 0)$.

By the same way as in the proof of Theorem 1, for each non-negative integer j, we can define the simple analytic curve $C_j : z = z_j(t)$ $(0 \leqq t < r_j \leqq +\infty)$ satisfying the following conditions:

$$z_j(0) = b_j, \qquad \operatorname{Im} z_j(t) > 0 \qquad (0 < t < r_j)$$

$$\lim_{t\to r_j} z_j(t) = \infty$$

and

$$f(z_j(t)) = f(b_j) + (-1)^{j+1}t \tag{5.6}$$

for $0 \leqq t < r_j$. Here if $a_{j-1} < a_j = \dots = a_{j+m-1} < a_{j+m}$ for some integers j and m with $m \geqq 2$, we further add the condition

$$\lim_{t \to +0} \arg(z_n(t) - z_n(0)) = \frac{m-1-n+j}{m}\pi$$

for $n=j,\dots,j+m-2$. Then these curves C_j are mutually disjoint and we obtain the sequence of the simply connected subregions D_j of the upper half plane H, each of which is bounded by C_j, C_{j+1}, and the segment b_jb_{j+1} of the real axis. Clearly from the construction of the curves C_j, any two regions D_m and D_n have no common points. In addition to these facts, from (2.1) and (5.6), each function $\mathrm{Im}\, z_j(t)$ is strictly increasing for $0 \leqq t < r_j$.

Let us consider the continuous function defined by

$$h(t) = \sum_{n \geqq 0} \int_0^t \frac{y}{(t-a_n)^2 + y^2}\, dt$$

for real values of t, where y is an arbitrarily fixed positive real number. Then h(t) is strictly increasing and unbounded above since all the a_n are positive. Further, it is clear from (2.1) that

$$h(t) = \arg f(iy) - \arg f(t+iy)$$

for real values of t. Hence for any sufficiently large r, it is possible to find a point of the half line $\{t+iy : t>r\}$ at which f(z) is real. From this fact and from (5.3), we can prove the next lemma by the same method as in the proof of Theorem 1.

LEMMA 5. *Each region* D_J *defined above does not contain any half line of the form* $\{t+iy : t>r\}$. *In particular, if* $\mathrm{Im}\, z_n(t)$ *is bounded above for some* n, *then*

$$\lim_{t \to r_n} \mathrm{Re}\, z_n(t) = -\infty$$

Assume now that all the curves C_j lie in some strip $0 \leqq \mathrm{Im}\, z \leqq M$. Then by this lemma, we have

$$\lim_{t \to r_j} \mathrm{Re}\, z_j(t) = -\infty$$

for every $j \geqq 0$. Therefore since $b_j = z_j(0) > 0$, all the curves C_j surely intersect the imaginary axis. Hence we can take a sequence $\{z_j\}$ such that $z_j \varepsilon C_j$, Re $z_j = 0$, and Im $z_j <$ Im $z_{j+1} < M$. Clearly this sequence converges to some finite point of the upper half plane H. However by (5.6), $(-1)^{j+1} f(z_j) > 0$ for $j \geqq 0$, so that $f(z)$ varishes at this limit point. This is impossible. Consequently for any positive number y, some curve C_n meets the line Im $z = y$.

Let y^* be an arbitrary positive real number. Then by the nature of the curves C_j, there exists a natural number m such that each C_n with $n \geqq m$ meets the line Im $z = y^*$ at exactly one point $x_n + iy^*$. Evidently $x_n < x_{n+1}$ and the intersection of D_n and this line is the open segment between the points $x_n + iy^*$ and $x_{n+1} + iy^*$. Furthermore, we remark that the x_n tends to infinity as n does to infinity. Besides, by making use of (2.1), we can see that Im $f(t+iy^*) \neq 0$ for each open interval (x_n, x_{n+1}). Hence $f(z)$ does not take any real values in each region D_n, so that the image $f(D_n)$ is contained entirely in the upper half plane or the lower half plane. It thus follows from (5.6) that the function $f(z)$ converges to infinity as z tends to infinity along each curve C_j, so that all the quantities r_j must be $+\infty$.

Here let us recall the monodromy theorem again. Then we obtain the following fact.

LEMMA 6. *Let* D_n *be the simply connected regions defined above. Then the function* $f(z)$ *maps each region* D_n *conformally onto the upper half plane or the lower half plane according to whether* n *is even or odd.*

5.4 Let G be the simply connected subregion of H whose boundaries consist of the curve C_0 and the half line $x \leqq b_0$ of the real axis. Then by what is mentioned just above, a point of H which does not belong to G must be contained in some C_j or D_j.

Assume now that there is a point s of G such that $f(s)$ is real positive. Then we can define the simple curve $C : z = z(t)$ $(0 \leqq t < r)$ satisfying $z(0) = s$ and

$$f(z(t)) = f(s) - t$$

for $0 \leqq t < r$. Since $f(z) \neq 0$ in H and $f'(z)$ vanishes only at the points b_j, this curve C meets neither C_0 nor the real axis. Hence C lies in the region G. Therefore C reaches to the point at infinity and along which $f(z)$ is real positive and bounded. On the other hand, by virtue of (2.1),

Im z(t) is decreasing for $0 \leqq t < r$. Hereby the curve C is contained entirely in the intersection of G and the strip $0 \leqq \text{Im } z \leqq \text{Im } s$. However, this is clearly impossible by (5.3). Hence the function f(z) does not assume any positive real values in the region G. Similarly f(z) has no negative real values in G. Accordingly, the image f(G) is contained in the lower half plane. Furthermore, we can see that f(z) maps G conformally onto the lower half plane. Consequently in the upper half plane H, f(z) takes real values only on the curves C_j. Hence from (5.6), we have the desired result at once. This completes the proof.

6. EXAMPLES

Hereafter for the sake of simplicity, we say that a real entire function f(z) satisfies A-condition if the set E(f) has the point 0 as its interior point.

Let us consider the function defined with

$$f_{1/s}(z) = \prod_{n\geqq 1} \left(1 - \frac{z}{n^s}\right) \tag{6.1}$$

where s is a real number greater than one. Then this function is real entire and its order is just 1/s. In [1], Bieberbach showed that if $s > 2$, then the function $f_{1/s}(z)$ satisfies A-condition. Furthermore, he proved that the function

$$f_0(z) = \prod_{n\geqq 1} \left(1 - \frac{z}{e^n}\right)$$

also fulfills A-condition.

In this final section we shall construct more functions which satisfy A-condition.

6.1 Let s be an arbitrary real number with $s > 1$ and let c_n be $(n+\frac{1}{2})^s$ for every natural number n. Then for the function defined by (6.1), we can easily find that

$$\liminf_{n\to+\infty} \frac{1}{n} \log|f_{1/s}(c_n)| \geqq \cot\frac{\pi}{s} \tag{6.2}$$

$$\liminf_{n\to+\infty} \frac{1}{n} \log|f_{1/s}(-c_n)| \geqq \pi\operatorname{cosec}\frac{\pi}{s} \tag{6.3}$$

Now let us consider the function defined by

$$F_{1/s}(z) = z\, f_{1/s}(z)\, f_{1/s}(-z) \tag{6.4}$$

Then it is clear from (6.2) and (6.3) that

$$\liminf_{n\to+\infty} \frac{1}{n} \log|F_{1/s}(c_n)| \geqq \pi(1 + \cos\frac{\pi}{s}) \operatorname{cosec}\frac{\pi}{s} \tag{6.5}$$

Hence we find that

$$\inf_{n\geqq 1} |F_{1/s}(c_n)| > 0 \tag{6.6}$$

by virtue of the above (6.5). On the other hand, it is clear that $F_{1/s}(-z) = -F_{1/s}(z)$ and $n^s < c_n < (n+1)^s$ for $n \geqq 1$. Therefore from Corollary, we can see at a glance that this function $F_{1/s}(z)$ surely satisfies A-condition.

Again let s be an arbitrary real number with $s > 1$, and let $\{a_n\}$ and $\{b_n\}$ be sequences of positive real numbers satisfying

$$\begin{aligned} (n - t_n)^s &\leqq a_n \leqq (n + t_n)^s \\ (n - t_n)^s &\leqq b_n \leqq (n + t_n)^s \end{aligned} \tag{6.7}$$

for every natural number n. Here $\{t_n\}$ is a sequence such that

$$0 < t_n < \frac{1}{2}, \qquad \sum_{n\geqq 1} t_n < +\infty \tag{6.8}$$

Then the orders of these sequences $\{a_n\}$ and $\{b_n\}$ are equal to 1/s, so that the canonical products

$$g(z) = \prod_{n\geqq 1} \left(1 - \frac{z}{a_n}\right) \tag{6.9}$$

$$h(z) = \prod_{n\geqq 1} \left(1 + \frac{z}{b_n}\right) \tag{6.10}$$

are both well defined. Now we further set

$$G(z) = z\, g(z)\, h(z). \tag{6.11}$$

Then this function G(z) is real entire and of order 1/s. In what follows we want to show that the function G(z) fulfills A-condition. To this end we compare G(z) with the function $F_{1/s}(z)$ defined by (6.4).

For each $n \geqq 1$, let us set

$$S_n(z) = \frac{n^s\,(a_n - z)}{a_n\,(n^s - z)} \tag{6.12}$$

Then it is clear from (6.1), (6.9), and (6.12) that

$$\frac{g(z)}{f_{1/s}(z)} = \prod_{n \geq 1} S_n(z) \tag{6.13}$$

for values of z. Furthermore, from (6.7) and (6.8) it is also clear that $a_n < c_n < a_{n+1}$ for every $n \geqq 1$. Hence by a simple estimation, we have

$$S_n(c_j) \geqq A_n,$$
$$A_n = \frac{n^s}{(n-t_n)^s} \frac{(n-t_n)^s - c_{n-1}}{n^s - c_{n-1}} \tag{6.14}$$

for $1 \leqq j < n$, and

$$S_n(c_j) \geqq B_n,$$
$$B_n = \frac{n^s}{(n+t_n)^s} \frac{(n+t_n)^s - c_n}{n^s - c_n} \tag{6.15}$$

for $n \geqq j$. Here let us note the inequalities

$$2^{1-s} sx \leqq 1 - (1-x)^s \leqq sx$$

for real values of x with $0 \leqq x \leqq 1/2$. Then from (6.14),

$$0 < 1 - A_n = \frac{(n-\tfrac{1}{2})^s}{(n-t_n)^s} \frac{n^s - (n-t_n)^s}{n^s - (n-\tfrac{1}{2})^s} \leqq 4^s t_n$$

for $n \geqq 1$. Similarly since

$$sx \leqq (1+x)^s - 1 \leqq 2^{s-1} sx$$

for $0 \leqq x \leqq 1/2$, it follows from (6.15) that

$$0 < 1 - B_n \leqq 4^s t_n$$

for $n \geqq 1$. Therefore by setting $C_n = \min(A_n, B_n)$, we can see that $0 < C_n < 1$ and

$$\prod_{n \geqq 1} C_n = M > 0$$

from (6.8). Besides, by (6.14) and (6.15), we have $S_n(c_j) \geqq C_n$ for any natural numbers n and j. It thus follows from (6.13) and (6.16) that

$$|g(c_n)| \geqq M \, |f_{1/s}(c_n)|$$

for $n \geqq 1$. On the other hand, by (6.7), (6.8), and (6.10), we can easily see that

$$\inf_{x \geqq 0} \frac{h(x)}{f_{1/s}(-x)} > 0$$

Therefore, by these facts, it is possible to find a positive real constant M^* such that

$$|G(c_n)| \geqq M^* \, |F_{1/s}(c_n)| \tag{6.17}$$

for each $n \geqq 1$. Furthermore, by the symmetry we can also find a positive constant M_* satisfying

$$|G(-c_n)| \geqq M_* \, |F_{1/s}(-c_n)| \tag{6.18}$$

for every natural number n. Consequently from (6.6) and Corollary, the set E(G) of this function G(z) surely contains the point 0 as its interior point. Hence G(z) fulfills A-condition. In addition to these facts, from (6.5), (6.17), and (6.18), the function G(z) also satisfies

$$\liminf_{n \to +\infty} \frac{1}{n} \log|G(c_n)| > 0$$

$$\liminf_{n \to +\infty} \frac{1}{n} \log|G(-c_n)| > 0$$

Hence for any real entire function K(z) satisfying $T(r,K) = o(r^{1/s})$, it is clear that

$$\inf_{n \geqq 1} \left| \frac{G(c_n)}{K(c_n)} \right| > 0, \qquad \inf_{n \geqq 1} \left| \frac{G(-c_n)}{K(-c_n)} \right| > 0$$

Hereby if the quotient G(z)/K(z) has no poles, then G(z)/K(z) also satisfies A-condition by means of Corollary.

6.2 We shall further present real entire functions of order one which satisfy A-condition.

Let t be an arbitrary real number with $0 < t < 1$, and for each natural number n, let a_n and b_n be any real numbers with $n+t \leqq a_n \leqq n+1$ and $-n-1 \leqq b_n \leqq -n-t$. Then the function defined by

$$W(z) = \prod_{n \geqq 1} \left(1 - \frac{z}{a_n}\right)\left(1 - \frac{z}{b_n}\right)$$

is real entire of order one and vanishes only at the points a_n and b_n. By an elementary analysis, we obtain

$$|W(x)| \geqq \left| \frac{(x-a_n) \sin\pi x}{a_n \, x(x^2-1)} \right|$$

for real values of x with $n < x < n+1$, and

$$|W(x)| \geqq \left| \frac{(x-a_n) \sin\pi x}{b_n \, x(x^2-1)} \right|$$

for $-n-1 < x < -n$ $(n \geqq 1)$. Hence for each natural number n, by setting $x_n = n + t/2$, we have

$$|W(x_n)| \geqq \frac{C^*}{a_n x_n (x_n^2-1)} \tag{6.19}$$

$$C^* = \frac{t}{2\pi} \sin \frac{t}{2} \pi$$

Similarly by setting $x_{-n} = -x_n$, we also have

$$|W(x_{-n})| \geqq \frac{C^*}{b_n x_{-n} (x_{-n}^2-1)} \tag{6.20}$$

for every $n \geqq 1$.

Now let us consider the real entire function defined with

$$f(z) = P(z)W(z)$$

where P(z) is a real polynomial of degree at least four so that all the zeros of f(z) are real and simple. Then by virtue of (6.19) and (6.20), it follows that

$$\inf_{n \neq 0} |f(x_n)| > 0$$

where the infimum is taken over all integers n except for 0. On the other hand, it is clear from the construction that

$$x_n < a_n < x_{n+1} < a_{n+1}$$

$$b_{n+1} < x_{-n-1} < b_n < x_{-n}$$

for each $n \geqq 1$. Therefore from Corollary, this real entire function f(z) satisfies A-condition.

Furthermore, it is also clear that

$$0 < a_{n+1} - a_n < 2, \qquad 0 < b_n - b_{n+1} < 2$$

for every $n \geqq 1$. Hence the function f(z) surely fulfills the assumptions of Theorem 5. Hereby the derivatives $f^{(n)}(z)$ of all orders must satisfy A-condition.

REFERENCES

1. L. Bieberbach, Über eine Vertiefung des Picardschen Satzes bei ganzen Funktionen endlicher Ordnung, *Math. Zeitschrift, 3* (1919), 175-190.

2. A. Edrei, Meromorphic functions with three radially distributed values, *Trans. Amer. Math. Soc., 78* (1955), 276-293.

VALUE DISTRIBUTION AND RELATED QUESTIONS IN ITERATION THEORY

L. S. O. Liverpool

Department of Mathematics
Fourah Bay College
University of Sierra Leone
Freetown, Sierra Leone

1. INTRODUCTION

The theory of iteration of rational or entire function $f(z)$ of the complex variable z deals with the sequence of iterates $f_n(z)$ defined by

$$f_1(z) = f(z), \quad f_{n+1}(z) = f_1(f_n(z)) \qquad n = 1, 2, 3, \ldots$$

In the theory developed by Fatou [9] and Julia [10], an important part is played by the set $F = F(f)$ of those points of the complex plane where $\{f_n(z)\}$ is not a normal family. Unless $f(z)$ is a rational function of order 0 or 1, the set $F(f)$ has the following properties:

(I) $F(f)$ is a non empty perfect set.

(II) $F(f_n) = F(f)$ for any integer $n \geq 1$.

(III) $F(f)$ is completely invariant under the mapping $z \to f(z)$, i.e., if α belongs to $F(f)$, then so do $f(\alpha)$ and every solution β of $f(\beta) = \alpha$.

(IV) For every $\alpha \varepsilon F(f)$ and for every complex β (excluding at most two exceptional β-values), there exists a sequence of positive integers $\{n_k\}$, k = 1, 2, ... and a sequence of complex numbers $\{\alpha_k\}$ such that

$$f_{n_k}(\alpha_k) = \beta \quad \text{and} \quad \lim_{k\to\infty} \alpha_k = \alpha$$

A fix point α of order n of f(z) is a solution of $f_n(z) = z$. It has exact order n if $f_k(\alpha) = \alpha$ for $k = n$, but not for any $k < n$ and in this case $f_n'(\alpha)$ is called the multiplier of α provided $\alpha \neq \infty$. The fix point is called attractive, indifferent, or repulsive according as the modulus of its multiplier is less than, equal to, or greater than 1, respectively. With this nomenclature we have:

(V) F(f) contains every repulsive fix point of f of any order. Attractive fix points belong to the complement of F(f).

(VI) F(f) is the derivative of the set of fix points (of all orders) of f. Indeed, it has been shown that it is the derivative of the set of repulsive fix points.

The complement C = C(f) of F(f) is an open set, whose components of connectivity are the maximal domains of normality of $\{f_n(z)\}$. By (III), C(f) is completely invariant. One of the main problems of the theory is to investigate the various possibilities for the structure of F(f) and the way in which F(f) divides the plane. Results on this problem have been obtained in the original papers of Fatou and Julia and several other authors.

It follows from (III) and (IV) that if F(f) has a non empty interior, then it is the whole plane. In the case of rational f, this can occur, as was shown by Lattes [12] in the case where f is the rational function such that P(2u) = f(P(u)), P(u) being the Weierstrass elliptic function. Baker [3] has now solved the long-standing open problem proposed by Fatou of finding an entire transcendental f for which F(f) is the whole plane.

P. Bhattacharyya [6] has studied the growth of entire functions with infinite (i.e., unbounded) domains of normality, and the author in [14] made an investigation of the restrictions placed on f by assuming F(f) and C(f) have certain properties.

The interaction between value distribution and iteration theory is a very natural one. In some cases, results in iteration can suggest the existence of results in value distribution. For example, since every point of F(f) is a point of accumulation of fix points of f(z) and also since for entire transcendental f, F(f) cannot be contained in any finite set of straight lines, one can conclude that if f(z) is entire transcendental and ℓ is any straight line in the complex plane, then F(f) is not restricted to

ℓ, and as such there is a positive integer k, for which $f_k(z) - z = 0$ has solutions not contained in ℓ. The obvious question becomes "what is the smallest k for which $f_k(z) - z = 0$ has solutions outside ℓ?" Since it is possible for all solutions of $f(z) - z = 0$, to lie on a line, by arbitrary prescription, the question becomes: Given an entire transcendental function f(z), and a straight line ℓ, in the complex plane, is it possible that $f(f(z)) - z = 0$ has solutions, which do not lie on ℓ? - a problem in value distribution! This question has been resolved for functions of order less than 1/2, but in general remains open.

On the other hand, work in value distribution frequently generates immediate results in Iteration theory. This is illustrated, for example, in [4] where methods and results, used to derive results in the value distribution of composite entire functions, are used to prove new results in Iteration theory and also in [8].

In this paper, we present a collection of recent results linking value distribution and Iteration theories.

2. STATEMENT OF RESULTS

Recently S. Kimura [11] proved

THEOREM A. Let f be an entire function of order less than one and w_n a sequence such that $|w_n| \to \infty$ as $n \to \infty$. Suppose that all the roots of the equations $f(z) = w_n$ $(n = 1, 2, \ldots)$ lie in a half-plane (say Re $z \geq 0$). Then f is a polynomial of degree at most 2.

We improve Theorem A slightly to

THEOREM 1. If f is an entire function whose growth is at most order one and minimal type, and w_n is a sequence such that $|w_n| \to \infty$ while all roots of $f(z) = w_n$ $(n = 1, 2, \ldots)$ lie in a half-plane, then f is a polynomial of degree at most 2.

In this form the theorem is sharp. For any $d > 0$, the function e^{dz} has type d and is bounded in Re $z \leq 0$ so that any sequence w_n such that $1 < |w_n| \to \infty$ may be taken to satisfy the hypothesis in Re $z \geq 0$. We shall present here also applications of Theorem 1, a result in value distribution to the theory of iteration of entire functions.

Sometimes it may happen that a component of $C(f)$ contains a half-plane. Thus as an example, for $d > 0$, the function

$$g(z) = d^{-1}(e^{dz} - 1) \tag{1}$$

maps $H = \{z: \text{Re } z < 0\}$ into itself so that $\{g_n\}$ is normal in H.

Suppose that conversely g is a transcendental entire function and that $C(g)$ contains a half-plane, which we may take to be Re $z < 0$. Then $F(g)$ lies in Re $z \geq 0$ and if we take a sequence $w_n \varepsilon F(g)$ such that $|w_n| \to \infty$, all solutions of $f(z) = w_n$ lie in (g) by the invariance property, and hence in Re $z \geq 0$. Thus from Theorem 1 we have

THEOREM 2. If g is transcendental entire function such that the domain of normality $C(g)$ of $\{g_n\}$ contains a half-plane, then the growth of g must be at least of order 1, positive type.

Example (1) shows that this is sharp with respect to growth. Related problems have been discussed under more restrictive conditions in [7].

If $0 \varepsilon F(g)$, then every solution z of $g(z) = 0$ belongs to $F(g)$. The following Theorem 3a is thus a strengthening of Theorem 2.

We introduce the notation

$$A(\theta,\delta) = \{z: |\arg z - \theta| < \delta\} \tag{2}$$

THEOREM 3a. Suppose (i) g is a transcendental entire function whose growth is at most of order 1, minimal type, (ii) all the zeros of g lie in Re $z \geq 0$. Then for any $\delta > 0$, the set $F(g) \cap A(\Pi,\delta)$ is unbounded.

Because of the importance of fixed points, it is interesting that we can also prove

THEOREM 3b. If in 3a (ii) is replaced by the hypothesis that the first order fixed points lie in Re $z \geq 0$, the conclusion remains true.

The example (1), for which all first order fixed points lie in Re $z \geq 0$, shows that 3b ceases to hold if the assumption of minimal type is dropped.

In the circumstances of Theorems 3a or 3b, it follows that $A(\Pi,\delta)$ must contain fixed points of some order of g. Can one be more explicit about the order of such fixed points? Let us take 3b and make the stronger hypothesis in (ii) that all the first order fixed points are real and positive. Our methods and results differ slightly according to the order of g. For order less than 1/2, we have

THEOREM 4a. Suppose (i) g is transcendental entire of at most order 1/2, minimal type, and (ii) all but finitely many first order fixed points of g are real and positive. Then for any $\delta > 0$, $A(\Pi,\delta)$ contains infinitely many fixed points of order k for each $k \geq 2$.

Indeed the fixed points of higher order, whose existence is shown in the theorem can be taken to be non-real. This is somewhat analogous to the result mentioned earlier that if f is transcendental entire of order less than 1/2 and ι is a straight line, then not all solutions of $f_2(z) - z = 0$ lie in ι. Neither result includes the other, but both show that second order fixed points tend to be scattered in their angular distribution.

If the order of g exceeds 1/2, we have not been able to prove the existence of fixed points of order 2 in $A(\Pi,\delta)$. However, we can prove

THEOREM 4b. If in Theorem 4a (i) is replaced by the assumption that the order of g is strictly positive, but at most order 1 minimal type, then for any θ,δ subject to $\frac{\Pi}{2} < \theta < \frac{3\Pi}{2}$, $\delta > 0$, we have that $A(\theta,\delta)$ contains infinitely many fixed points of order k for each $k \geq 3$.

Thus in particular if g is at most of order 1 minimal type and all first order fixed points are real and positive, f has fixed points of every order greater than 2 in $A(\Pi\ \delta)$, however small $\delta > 0$ is taken.

The arguments used in this discussion can also be applied to show that functions of certain classes are not expressible as iterates of entire functions. An example is furnished by

THEOREM 5. Suppose the transcendental function F is such that

(i) $\limsup_{r\to\infty}\{\log\log\log M(F,r)\}/\log r < 1$,

(ii) all first order fixed points of F lie in $\mathrm{Re}\, z \geq 0$, and

(iii) F is bounded in $A(\Pi,\delta)$ for some $\delta > 0$.

Then F is not expressible as f_k, $k \geq 2$, for any entire f.

In (ii) we may replace fixed points by zeros without affecting the validity of the theorem. The function e^{e^z} has all its fixed points in $\operatorname{Re} z \geq 0$ and shows that we cannot allow equality in (i).

3. PROOF OF THEOREM 1

We may assume $f(0) \neq 0$ (otherwise consider $f(z - \delta)$ for a suitable positive constant δ).

We shall use the following results about functions of minimal type whose zeros lie in a half-plane. They may be found, e.g., in the proof of Theorem 1 of Liverpool [14] where the additional hypothesis $f(-r) = O(r^k)$ of that theorem is not used until after these facts have been derived.

LEMMA 1. Let f be a transcendental entire function of at most order one and minimal (i.e., zero) exponential type. Suppose $f(0) \neq 0$ and that all zeros a_n of f lie in the right half plane $\operatorname{Re} z \geq 0$. Then there are constants A and c such that

$$f(z) = Ae^{cz} \prod_{n=1}^{\infty} \left(1 - \frac{z}{a_n}\right) e^{\frac{z}{a_n}} \tag{3}$$

where $a_n = r_n e^{i\theta_n}$ is such that

$$\lambda = \operatorname{Re} \sum_{n=1}^{\infty} a_n^{-1} = \sum_{n=1}^{\infty} (\cos \theta_n)/r_n \tag{4}$$

is convergent and

$$\lambda + \operatorname{Re} c = 0 \tag{5}$$

Further, for any fixed k

$$|f(-r)|/r^k \to \infty \text{ as } r \to \infty \tag{6}$$

Proof of Theorem 1. We may suppose $w_1 = 0$ (for otherwise consider $f(z) - w_1$) and suppose first that f is transcendental entire of at most order one, minimal type and that all solutions of $f(z) = w_n$ lie in $H: \operatorname{Re} z \geq 0$. In particular the zeros $a_n = r_n e^{i\theta_n}$ lie in H, so by Lemma 1

$$\frac{f'(z)}{f(z)} = c + \sum_{n=1}^{\infty} \left(\frac{1}{z - a_n} + \frac{1}{a_n}\right)$$

Using (4) and (5) this yields

$$\operatorname{Re}\frac{f'(z)}{f(z)} = \sum_{n=1}^{\infty} \operatorname{Re}\frac{1}{z - a_n} \tag{7}$$

If Re z < 0 and Re a $\geq$ 0, we have $\operatorname{Re}\frac{1}{z - a} < 0$, while if $z = \rho e^{i\phi}$, then for fixed ϕ

$$|z| \operatorname{Re}\frac{1}{z - a} \to \cos\phi \text{ as } \rho \to \infty$$

Thus by (7), if δ is a fixed number such that $0 < \delta < \frac{\Pi}{2}$,

$$|z| \operatorname{Re}\frac{f'(z)}{f(z)} \to -\infty \text{ as } z \to \infty \text{ in } A(\Pi,\delta)$$

Take a fixed constant $K > 2\Pi/\delta$. Then there is a constant r_0 such that

$$|z| \left|\frac{f'}{f}\right| > K \quad \text{for} \quad z \in A(\Pi,\delta),\ |z| > r_0 \tag{8}$$

Next choose a member of the given sequence w_n so that $|f(z)| < |w_n|$ for $|z| \leq r_0$. By (6) there is a largest r_n such that $|f(-r_n)| = |w_n|$. There is a component G of $\{z: |f(z)| > |w_n|\}$ which contains $\{z: z = -r < -r_n\}$ and this component is bounded by a level curve $\Gamma: |f(z)| = |w_n|$ which passes through $z = -r_n$. Γ cannot close in Re z < 0 for there are no zeros of f in this region.

If Γ meets neither of the lines arg $z = \Pi \pm \delta$, then G lies entirely in the angle $A(\Pi,\delta)$. Let $r\theta(r)$ be the length of that segment γ_r of $|z| = r$ which lies in G and contains $z = -r$. By the arguments used in the proof of the Denjoy-Carleman-Ahlfors theorem in [15, pp. 310-311], it follows that for all sufficiently large r ($>r_1$, say), the maximum modulus function M(f,r) of f satisfies

$$\log\log M(f,r) > \log\log \max_{\gamma_r} |f(z)| > \Pi \int_{r_1}^{r} \frac{dt}{t\theta(t)} + C$$

for a suitable constant C. Since $\theta(r) < 2\delta$ this implies that f has order at least $\Pi/2\delta > 1$, which is impossible.

Thus there is a level curve $\Gamma: |f(z)| = |w_n|$, which starts at $z = -r_n$ and runs to either arg $z = \Pi + \delta$ or $\Pi - \delta$. Moreover Γ lies in $|z| \geq r_0$ so that the inequality (8) holds on Γ. But $w = f(z)$ maps Γ onto $|w| = |w_n|$ and as z traverses Γ, w traverses $|w| = |w_n|$ without change of direction. Further, we have

$$\frac{dw}{w} = \frac{dz}{z}\frac{zf'(z)}{f(z)}$$

whence, if $w = |w_n|e^{i\phi}$ and $z = re^{i\theta}\varepsilon\Gamma$, we have

$$i\,d\phi = \left(\frac{dr}{r} + i\,d\theta\right)\frac{zf'(z)}{f(z)} \tag{9}$$

so that by (8)

$$|d\phi| \geq |d\theta|\left|\frac{zf'(z)}{f(z)}\right| > K|d\theta|$$

The image of Γ is therefore an arc of $|w| = |w_n|$ whose angular measure is at least $K\delta > 2\Pi$. Thus Γ, and in particular $A(\Pi,\delta)$ must contain a root z of $f(z) = w_n$, against the hypothesis of the theorem.

We conclude that f cannot therefore be transcendental. If f is a polynomial, its degree can clearly not exceed two.

4. PROOF OF THEOREM 3

Suppose g is a transcendental entire function of growth at most order 1, minimal type and is such that

$$A(\Pi,\delta)\cap F(g) \tag{10}$$

is bounded for some $\delta > 0$. Without loss of generality, we may assume that the set in (10) is empty - it is only necessary to shift the origin and consider the iteration of $g(z + a) - a$ for sufficiently large negative a.

Whether the zeros of g(z) or the fixed points (i.e., the zeros of $g(z) - z$) are in Re $z \geq 0$, it follows from Lemma 1 that for any k

$$g(-r)/r^k \to \infty \text{ as } r \to \infty \tag{11}$$

Since $A = A(\Pi,\delta)$ does not meet F, A belongs to an unbounded component G of the set C(g) of normality of g^n. Indeed by [5], G is simply connected. The boundary ∂G belongs to F and is a continuum in the complex sphere. By the invariance property of F, g(z) omits all the values of ∂G for $z\varepsilon A$.

If $M = \Pi/(2\delta)$, the transformation T:

$$u = (1 + t)/(1 - t), \quad z = -u^{1/M}$$

maps $|t| < 1$ onto A, so that the function

$$w = h(t) = g\left\{-\left(\frac{1 + t}{1 - t}\right)^{1/M}\right\}$$

is regular in $|t| < 1$ and omits the values $w\varepsilon\partial G$.

By a result of J. E. Littlewood [13]

$$M(h,\rho) = O\{1 - \rho)^{-2} \quad \text{as} \quad \rho \to 1^-$$

If $z = re^{i\theta} \varepsilon A$, and $|\theta - \Pi| < \delta/2$, then in T

$$t = (1 - e^{Mi(\pi-\theta)}r^{-M})/(1 + e^{Mi(\pi-\theta)}r^{-M})$$

$$\sim 1 - 2e^{Mi(\pi-\theta)}r^{-M} \quad \text{as} \quad r \to \infty$$

Since $|M(\Pi - \theta)| < \Pi/4$, we have $1 - |t| > r^{-M}$ for large r. Thus as $z = re^{i\theta} \to \infty$ in $|\theta - \Pi| < 1/2\delta$, we have

$$|g(z)| = |h(t)| < M(h,1 - r^{-M}) = O(r^{2M})$$

But this conflicts with (11). The result follows.

5. PRELIMINARIES TO THE PROOF OF THEOREMS 4a AND 4b

Throughout this section assume that g is an entire function such that

(i) g is transcendental and of at most order one, minimal type,

(ii) all but finitely many fixed points of first order of g are real and positive. Then we have

$$g(z) - z = p(z)e^{cz} \prod_{n=1}^{\infty} \left(1 - \frac{z}{a_n}\right)e^{\frac{z}{a_n}}$$

where p is a polynomial of degree say $d \geq 0$, and $a_n > 0$. Applying Lemma 1 to $\{g(z) - z\}/p(z)$, we see that Σa_n^{-1} converges and in fact

$$g(z) - z = p(z) \exp(i\gamma z) \prod_{n=1}^{\infty} (1 - \frac{z}{a_n})$$

where γ is real. If $\gamma \neq 0$, then

$$\text{Max } |g(\pm iy)| > \exp |\gamma y|$$

so $\gamma = 0$ since g has minimal type. Thus

$$g(z) = z + h(z), \ h(z) = p(z) \ Q(z) = p(z) \prod_{n=1}^{\infty} (1 - \frac{z}{a_n}) \tag{12}$$

LEMMA 2. If g satisfies (i),(ii), then there is some $r_o > 0$ such that $|g(-r)|$ is increasing for $r > r_o$, so that $w = g(-r)$, $r > r_o$ describes a simple curve Γ. Γ approaches infinity in a limiting direction $\arg w = \alpha$.

For, let δ satisfy $0 < \delta < \frac{\Pi}{2}$. From (12) it follows that as $z \to \infty$ in $A(\Pi,\delta)$, we have $|h(z)/z| \to \infty$ and

$$\left|\frac{zh'}{h}\right| = \left|\frac{zp'}{p} + \frac{zQ'}{Q}\right| = \left|d + o(1) + \frac{zQ'}{Q}\right| \to \infty$$

(c.f. (7) and (8) in Theorem 1). Thus $|h'(z)| \to \infty$ and

$$\frac{zg'}{g} = \frac{zh'}{h}\frac{(1 + 1/h')}{(1 + z/h)} \to \infty \quad \text{as } z \to \infty \text{ in } A(\Pi,\delta) \tag{13}$$

In particular

$$g'(-r)/g(-r) = \frac{-1}{r}\{d + o(1) + \sum_{n=1}^{\infty} \frac{r}{r + a_n}\}\{1 + o(1)\} \tag{14}$$

as $r \to \infty$, and if $g(-r) = Re^{i\phi}$, we have

$$\frac{dR}{R} + id\phi = \frac{g'(-r)}{g(-r)}(-dr) \tag{15}$$

By (14) the argument of (15) approaches zero as $r \to \infty$, so that $\frac{dR}{R} > 0$ for larger r.

Clearly $|h(-r)| \to \infty$ faster than any power of r and arg $h(-r)$ tends to a constant value, namely the argument of the leading coefficient of $p(z)$. Hence arg $g(-r)$ approaches the same limit. The lemma is proved.

LEMMA 3. If g satisfies (i),(ii) then, given any real θ_0,δ,σ such that $0 < \delta < \frac{\Pi}{2}$, $0 < \sigma \leq \Pi$, there exist a constant R_1 and two branches Ψ and χ of $z = g^{-1}(w)$ regular in

$$S = A(\theta_0,\sigma) \quad \{|w| > R_1\}$$

such that the values of Ψ, χ satisfy $\Pi - \delta < \arg \Psi < \Pi$ and $\Pi < \arg \chi < \Pi + \delta$, respectively. For any $k > 0$ we have

$$\text{Max}\{|\Psi(w)|,|\chi(w)|\} = O(|w|^{1/k}) \quad \text{as} \quad w \to \infty \text{ in } S \tag{16}$$

Proof. As w traverses Γ from $w_0 = g(-r_0)$ to ∞ the branch of $z = g^{-1}(w)$ such that $r_0 = g^{-1}(w_0)$ has a regular continuation and the values of z are all real and negative ($< -r_0$).

For $r_1 > r_0$ put $R = |g(-r_1)|$ and consider the level-curve $\lambda = |g(z)| = R$ which passes through $z = -r_1$. Along λ we have as in (9)

$$id\phi = (zg'/g)\{id\ + \frac{dr}{r}\}$$

where $z = re^{i\theta} \varepsilon\lambda$, $g(z) = Re^{i\phi}$.

By (13) for z of suggiciently large modulus in $A(\Pi,\delta)$ we have for any given $K > 4\Pi/\delta$ that $|zg'(z)/g(z)| > K$. Thus if R and hence r are sufficiently large we have $|d\phi| > K|d\theta|$, $|d\phi| > K|dr|/r$. As z leaves $-r_1$ on λ and travels in a given direction to $re^{i\theta}$, the corresponding changes monotonely so that

$$K|\theta - \Pi| = |\int K d\theta| < \int K|d\theta| \leq \int|d\phi| = |\int d\phi| = \Delta\phi$$

and similarly $K|\log(r/r_1)| \leq \Delta\phi$. As $w = g(z)$ traverses $|w| = R$, increasing from arg $g(-r)$ by 4Π, z traverses λ in one direction with θ changing by at most $4\Pi/K < \delta$, while r satisfies

$$r_1\exp(-4\Pi/K) < r < r_1 \exp(4\Pi/K) \tag{17}$$

Thus if r_1 is large enough, the value of z remains in $A(\Pi,\delta)$ and by (13) $g'(z) \neq 0$ on λ so the value of z gives a regular continuation of $g^{-1}(w)$ from $g(-r_1)$ in Γ round $|w| = R$ through an angle of 4Π. The values of z lie in $A(\Pi,\delta)$ but do not meet the negative real axis except at $z = -r_1$, since $g(-r)$ is increasing. Since can be taken to lie in any sector $|\arg w - \alpha| < \varepsilon$, $\varepsilon > 0$, it follows that we can derive from thses values of $g^{-1}(w)$ a branch Ψ which satisfies the statements of Lemma 3, including either $\Pi - \delta < \arg \Psi < \Pi$ or $\Pi < \arg \Psi < \Pi + \delta$.

If in the above construction we begin by proceeding along λ in the opposite direction from that chosen originally, we construct the other branch χ of g^{-1}.

For $re^{i\theta} = \Psi(Re^{i\phi})$ we have by

$$|r| = |\Psi(Re^{i\phi})| < r_1 \exp(4\Pi/K)$$

and from $R = |g(-r_1)| > r_1^{2k}$ for large r_1 the estimate (16) follows.

We shall also need

LEMMA 4 (Polya [16]). Let e,f,h be entire functions which satisfy $e = foh$, $h(0) = 0$. Then there is a positive constant c independent of e,f,h such that

$$M(e,r) > M\left[g, cM\left(h, \frac{r}{2}\right)\right] \tag{18}$$

The condition $h(0) = 0$ can be dropped if (18) is to hold only for all sufficiently larger r.

6. PROOFS OF THEOREMS 4a AND 4b

THEOREM 4a. Suppose g satisfies the hypotheses of the theorem. The first of these implies that the minimum modulus of g is large ($> R_n$) on a sequence of circles $|z| = R_n \to \infty$. The R_n may be chosen so that there is at least one zero of g in each $R_n < |z| < R_{n+1}$. Since $|g(-r)|/r \to \infty$ as $r \to \infty$, each of the simply connected slit annuli

$$A_n = \{z: R_n < |z| < R_{n+1}, |\arg z| < \Pi\}, \quad n = 1, 2, \ldots$$

contains a zero of g and has the property that

$$|g(z)| > |z| \quad \text{on the boundary } \partial A_n \tag{19}$$

Denote by ϕ a branch of $z = g^{-1}(w)$ which is regular in $A(0,\Pi)$ for sufficiently large w, with values in $\Pi > \arg z \; \Pi \; \Pi - \delta$, δ being the fixed number, $0 < \delta < \frac{\Pi}{2}$ chosen in 5. Such a ϕ exists by Lemma 3.

For any fixed $\iota = 2, 3, \ldots$, the $(\iota - 1)$-th iterate $\phi_{\iota-1}(w)$ is defined in $A(0,\Pi)$ for sufficiently large w, with values in $\Pi > \arg z > \Pi - \delta$. For sufficiently large n then $\phi_{\iota-1}$ maps A_n univalently onto a simply connected domain D_n in $\Pi > \arg z > \Pi - \delta$. For $z \varepsilon \partial D_n$ we have $g_{\iota-1}(z) \varepsilon \partial A_n$. Now since $|g(z)| > |z|$ for large $|z|$, $z \varepsilon A(\Pi,\delta)$, it follows from $z \varepsilon \partial D_n$ that $|g_{\iota-1}(z)| > |z|$ and from $g_{\iota-1}(z) \varepsilon \partial A_n$ and (19) that at

$$|g_\iota(z)| = |g(g_{\iota-1}(z))| > |g_{\iota-1}(z)| > |z|$$

at least for large n.

By Rouche's theorem, $g_\iota(z) - z$ and $g_\iota(z)$ have equal numbers of zeros in D_n and $0 \varepsilon g(A_n) = g_\iota(D_n)$. Thus the region $\Pi > \arg z > \Pi - \delta$ and a fortiori $A(\Pi,\delta)$ contains an infinity of solutions of $g_\iota(z) - z = 0$.

THEOREM 4b. Suppose g has order ρ, $0 < \rho \leq 1$, and is at most of order one, minimal type, while all but finitely many first order fixed points are positive. Suppose also that $\frac{\Pi}{2} < \theta < \frac{3\Pi}{2}$ and that σ, $0 < \sigma < \Pi/2$ is so small that $\frac{\Pi}{2} < \theta \pm \sigma \leq \frac{3\Pi}{2}$. Let Ψ and χ be the two branches of g^{-1} whose existence is asserted in Lemma 3, in the case $\theta_o = \theta$. Then $\Psi = \chi$ has no solution in $A(\theta,\sigma) \cap \{|w| > R_1\}$.

Suppose g has only finitely many fixed points of order k in $A(\theta,\sigma)$. Then

$$F = (g_{k-1} - \Psi)/(\chi - \Psi)$$

is regular and different from 0, 1, ∞ for large z in $A(\theta,\sigma)$. By applying Schottky's theorem to F in $A(\theta,\sigma)$ (or in a slightly smaller sector within $A(\theta,\sigma)$ and with origin shifted so that $F \neq 0, 1, \infty$ in this sector) we find

$$F(z) = O\{\exp(C|z|^{\Pi/\sigma})\} \tag{20}$$

for some constant C as $z \to \infty$ in $A(\theta,\sigma')$, $\sigma' < \sigma$. From (16) the same estimate follows for $|g_{k-1}(z)|$ with perhaps a different C.

Now there exists δ_1 such that $0 < \delta_1 < \frac{\Pi}{2}$ and $A(\theta,\sigma')$ $A(\Pi,\delta_1)$. Thus $|g(re^{i\theta})| \to \infty$ as $r \to \infty$ and $|zg'/g| > K > 2\Pi/\sigma'$ for large $|z|$, $z\varepsilon A(\theta,\sigma')$. As in the proof of Theorem 1, there is for large R a level curve $\Gamma(R)$: $|g(z)| = R$, which passes through $z = re^{i\theta}$, say. Such a curve cannot close in $A(\theta,\delta)$ for arbitrarily large R, since $|g(z)| \to \infty$ in $A(\theta,\delta)$ and $A(\theta,\delta)$ contains only finitely many zeros of g. As in Theorem 1, Γ must run to the boundary of $A(\theta,\sigma')$ in at least one direction. If γ is an arc of Γ which goes from $re^{i\theta}$ to $\partial A(\theta,\sigma')$, then from $|zg'/g| > K$ it follows that the image of γ under $w = g(z)$ is the whole of $|w| = R$.

For large R we have that if t is the point on $|t| = R$ where $|g_{k-2}(t)| = M(g_{k-2},R)$, then for some $z\varepsilon\gamma$, $g(z) = t$

$$M(g_{k-2},R) = |g_{k-1}(z)| \tag{21}$$

Now in $A(\theta,\sigma') \subset A(\Pi,\delta_1)$, $|g(z)|/|z|^N \to \infty$ as $|z| \to \infty$, for any N. Take $N > 2\Pi/(\rho\sigma)$, where ρ is the order of g. Then for large R we have from (21)

$$\max_{\substack{|z|=r \\ z\varepsilon A(\theta,\sigma')}} |g_{k-1}(z)| > M(g_{k-2},r^N) \tag{22}$$

Since $k - 2 \geq 1$, the right hand side is (for large r) at least

$$M(g,r^N) > \exp(r^{N\rho}) > \exp(r^{2\Pi/\sigma})$$

for some arbitrarily large r. Thus we have a contradiction between (22) and the estimate for g_{k-1} from (20). Hence g must in fact have an infinity of fixed points in $A(\theta,\sigma)$.

7. PROOF OF THEOREM 5

Suppose F satisfies the hypotheses of Theorem 5 and that there exist an entire function f and an integer $k \geq 2$ such that $F = f_k$. Since F is bounded on the path γ which consists of the negative axis running to $-\infty$, it follows that one of γ, $f(\gamma),\ldots,f_{k-1}(\gamma)$ is an unbounded path on which f is bounded. From this it follows that the lower order of f is positive.

From Lemma 4 and the fact that the lower order of f is positive, we easily obtain a contradiction to hypothesis (i) of the theorem, provided $k \geq 3$.

It remains to prove the theorem for $k = 2$. From hypothesis (i), $F = f_2$ and the fact that the lower order of f is positive, it follows from Lemma 4 (as is proved in [1, Satz 12] that the order of f is less than one.

Now $f(z) = z + g(z)$ where the zeros of g are fixed points of f and hence of F. Thus the zeros of g lie in Re $z \geq 0$ and the order of z is less than 1. By Lemma 1 we have

$$\frac{|f(-r)|}{r^2} \text{ and } \frac{|g(-r)|}{r^2} \to \infty \quad \text{as} \quad r \to \infty \tag{23}$$

while

$$|z|\left|\frac{g'}{g}\right| > K > 2\Pi/\delta \quad \text{in} \quad |z| > r_o, \ |\arg z - \Pi| < \delta \tag{24}$$

For a large $R(> M(g,r_o))$ there is a level curve Γ: $|g(z)| = R$ passing through $z = -r$ such that $|g(-r)| = R > r^2$. Just as in the proof of Theorem 1, it follows that Γ must run to at least one of $\arg z = \Pi + \delta$ or $\Pi - \delta$, say the former, and that the image under $w = g(z)$ of this arc must cover $|w| = R$ with angular measure at least $K\delta > 2\Pi$. Let γ denote the arc of Γ between $-r$ and a point z' chosen so that the image $g(\gamma)$ covers exactly the angular length $K\delta$ of $|w| = R$. As in the proof of Lemma 3 (17), it follows that for all $z_1 = r_1 e^{i\theta_1} \varepsilon \gamma$, we have $|\log(r_1/r)| < \delta$.

The arc γ is mapped by $f(z) = z + g(z)$ onto a (not necessarily closed) curve in such a way that the image of z_1 is $z_1 + Re^{i\phi_1}$ where $|z_1| < re^{\delta}$, $R > r^2$, and ϕ_1 increases by $K\delta > 2\Pi$ as z_1 describes γ. Thus $f(\gamma)$ certainly cuts the negative real axis, say in a point $w' = f(z'')$, $z'' \varepsilon \gamma$. Then

$$|F(z'')| = |f(f(z''))| = |f(w')| > |w'|^2 > (R - re^{\delta})^2 > 1/4\ r^4$$

if R and hence r are sufficiently large. Thus $A(\Pi,\delta)$ contains points z'' of arbitrarily large modulus for which

$$|F(z'')| > 1/4 e^{-4\delta} |z''|^4$$

which contradicts (iii). This completes the proof.

REFERENCES

1. I. N. Baker, Zusammensetzungen ganzer Funktionen, *Math. Z, 69* (1958), 121-163.

2. I. N. Baker, The distribution of fixed points of entire functions, *Proc. London Math. Soc. (3), 16* (1966), 493-506.

3. I. N. Baker, Limit functions and sets of non-normality in iteration theory, *Ann. Acad. Sci. Fenn. Ser. A.I Math., 467* (1970), 1-111.

4. I. N. Baker, The value distribution of composite entire functions, *Acta Sci. Math. Tom., 32* (1971) 87-90.

5. I. N. Baker, The domains of normality of an entire function, *Ann. Acad. Sci. Fenn. Ser. A.I., 1* (1975), 277-283.

6. P. Bhattacharyya, *Iteration of Analytic Functions*, Ph.D. Thesis, Univ. of London, 1969.

7. P. Bhattacharyya, On the domain of normality of an attractive fixed point, *Trans. Amer. Math. Soc., 153* (1971), 89-98.

8. I. N. Baker and L. S. O. Liverpool, The value distribution of entire functions of order at most one, *Acta. Sci. Math. Tom., 41* (1979), 3-14.

9. P. Fatou, Sur l'iteration des fonctions transcendantes entieres, *Acta. Math., 47* (1926), 337-370.

10. G. Julia, Memoire sur l'iteration des fonctions rationelles, *J. Math. pures Appl. (8) 1* (1918), 47-245.

11. S. Kimura, On the value distribution of entire functions of order less than one, *Kōdai Math. Sem. Rep. 28* (1976), 28-32.

12. S. Lattes, Sur l'iteration des substitutions rationnelles et les fonctions de Poincare, *C. R. Acad. Sci. Paris, 166* (1918), 26-28.

13. J. E. Littlewood, On inequalities in the theory of functions, *Proc. Lond. Math. Soc., 23* (1925), 481-519.

14. L. S. O. Liverpool, On entire functions with infinite domains of normality, *Aeq. Math., 10* (1974), 189-200.

15. R. Nevanlinna, *Eindeutige analytische Funktionen, 2 Aufl.*, Springer (Berlin, 1953).

16. G. Polya, On an integral function of an integral function, *J. Lond. Math. Soc., 1* (1926), 12-15.

ON ANALYTIC MAPPINGS AND FACTORIZING MEROMORPHIC FUNCTIONS

Kiyoshi Niino

Faculty of Technology
Kanazawa University
Kodatsuno, Kanazawa
Japan

1. FAMILY OF ANALYTIC MAPPINGS BETWEEN TWO ULTRAHYPERELLIPTIC SURFACES

Let R and S be two ultrahyperelliptic surfaces defined by two equations

$$y^2 = G(z) \quad \text{and} \quad y^2 = g(z) \tag{1}$$

respectively, where G(z) and g(z) are entire functions, each of which has an infinite number of simple zeros and no other zeros.

Ozawa [13] proved the following theorem, which connects problems of analytic mappings between two ultrahyperelliptic surfaces with problems of functional equations involving composite functions.

THEOREM A. *There exists a non-trivial analytic mapping* ψ *of* R *into* S, *if and only if there is a non-constant entire function* h(z) *satisfying the functional equation*

$$f(z)^2 G(z) = g(h(z)) \tag{2}$$

with a suitable entire function f(z).

The entire function h(z) in (2) is called the *projection* of the analytic mapping ψ. We denote by A(R, S) be the family of non-trivial analytic mappings of R into S and by H(R, S) the family of projections of analytic mappings in A(R, S). Let $H_P(R, S)$ be the subfamily of H(R, S) consisting of polynomials, and $H_T(R, S)$ the subfamily of H(R, S) consisting of transcendental entire functions.

We study the relation between different analytic mappings of a given R into a given S and find it, in terms of the proejctions. In [7] Mutō proved

THEOREM B. *If* $H(R, S) \neq 0$, *then* $H(R, S) = H_P(R, S)$ *or* $H(R, S) = H_T(R, S)$.

Many results on $H_P(R, S)$ have been obtained by Miromi-Mutō [6], the author [9], (10], Mutō and the author [8] and Baker [1]. With respect to $H_T(R, S)$, Mutō and the author [8] proved

THEOREM C. *Let* R *and* S *be two ultrahyperelliptic surfaces with* $P(R) = P(S) = 4$, *where* P(R) *and* P(S) *are the Picard constants of* R *and* S, *respectively* (cf. Ozawa [12]). *If* $H_T(R, S) \neq 0$, *then* $H_T(R, S)$ *consists of transcendental entire functions of the same order, the same type, and the same class.*

In this paper we shall consider $H_T(R, S)$ without the condition $P(R) = P(S) = 4$. Our results are the following:

THEOREM 1. *Let* R *and* S *be two ultrahyperelliptic surfaces defined by the equations* (1), *respectively. Suppose that* g(z) *is of finite order and positive lower order, and that there are positive numbers* c *and* r_0 *such that*

$$T(r, g) \leqq c\, N(r, 0, g) \tag{3}$$

hold for all $r \geqq r_0$. *If there are two different analytic mappings* ψ_1 *and* ψ_2 *of* R *into* S *whose projections* $h_1(z)$ *and* $h_2(z)$ *are of finite order, respectively, then the both orders are equal.*

THEOREM 2. *Let* R *and* S *be two ultrahyperelliptic surfaces defined by the equations* (1), *respectively. Suppose that* g(z) *is of finite order, of positive lower order and satisfies* (3). *If there is an analytic mapping* ψ *of* R *into* S *such that the projection* h(z) *of* ψ *is of finite order* λ_h, *then there is no analytic mapping of* R *into* S *such that its projection is of lower order greater than* λ_h.

We assume that the reader is familiar with the Nevanlinna theory of meromorphic functions and the usual notations such as $T(r, f)$, $M(r, f)$, $N(r, a, f)$, $n(r, a, f)$, etc. (see e.g. [5]).

2. FACTORIZING MEROMORPHIC FUNCTIONS

Baker-Gross [2] asked: under what circumstances can an entire function $F(z)$ have two factorizations

$$F(z) = f(p(z)) = f(q(z)) \tag{4}$$

where $p(z)$ and $q(z)$ are polynomials, and the left factor $f(z)$ is the same in each case? And they obtained the following:

THEOREM D. *If* $f(z)$ *is a non-constant entire function, and* $p(z)$, $q(z)$ *are non-constant polynomials satisfying* (4), *then either* (i) *there exist a root of unity* λ *and a constant* β *such that* $p(z) = \lambda q(z) + \beta$ *or* (ii) *there exist a polynomial* $r(z)$ *and constants* c, s *such that* $p(z) = r(z)^2 + s$, $q(z) = (r(z) + c)^2 + s$.

In case (i) *either* $\lambda = 1$ *and* $f(z)$ *is periodic with period* β, *or* λ *is a primitive* j-*th root of unity*, $j > 1$, *in which case* $f(z)$ *has the form*

$$f(z) = \sum_{n=0}^{\infty} a_n (z - \eta)^{nj}, \quad \eta = \beta/(1 - \lambda)$$

If $c \neq 0$ *in case* (ii), *then* $f(s + z^2)$ *is an even periodic function of period* c.

All the cases mentioned above do occur.

In the case when $p(z)$ and $q(z)$ are transcendental in (4), we can deduce from Lemma in Mutō and the author [8] that

THEOREM E. *If* $f(z)$ *is a non-constant entire function, and* $h(z)$, $k(z)$ *are two transcendental entire functions satisfying*

$$F(z) = f(h(z)) = f(k(z))$$

then $h(z)$ *and* $k(z)$ *are of the same order, the same type, and the same class.*

Now we shall consider the following problem: let $F(z)$ be a meromorphic function having two factorizations

$$F(z) = f(h(z)) = f(k(z)) \tag{5}$$

where $h(z)$ and $k(z)$ are entire, and the left factor $f(z)$ is the same meromorphic function in each case. What can be said about the relation between $h(z)$ and $k(z)$?

Firstly, we have the result of Baker-Gross [2], that is,

THEOREM F. *If* $F(z)$ *and* $f(z)$ *are non-constant meromorphic functions, and* $h(z)$, $k(z)$ *are non-constant polynomials satisfying* (5), *then* $h(z)$ *and* $k(z)$ *have the same degree, and the ratio of their leading coefficients has modulus* 1.

Professor M. Ozawa orally pointed out to the author that the above problem is closely connected with the problem of analytic mapping as follows:

Let $h(z)$ and $k(z)$ be two elements of $H(R, S)$ mentioned in Section 1. Then it follows from Theorem A that there are two entire functions $F_1(z)$ and $F_2(z)$ satisfying

$$F_1(z)^2 G(z) = g(h(z)) \quad \text{and} \quad F_2(z)^2 G(z) = g(k(z))$$

Hence we have the following functional equation

$$g(h(z)) = \alpha(z)^2 g(k(z)) \tag{6}$$

where $\alpha(z) = F_1(z)/F_2(z)$ is meromorphic and $g(z)$ is an entire function having an infinite number of simple zeros and no other zeros.

On the other hand, for a meromorphic function $f(z)$ we write

$$f(z) = \frac{f_1(z)^2 g_1(z)}{f_2(z)^2 g_2(z)} \tag{7}$$

where $f_j(z)$ and $g_j(z)$ $(j = 1, 2)$ are entire functions such that each $g_j(z)$ has simple zeros only and is the canonical product of least possible genus over these zeros, and $f_1(z)^2 g_1(z)$ and $f_2(z)^2 g_1(z)$ have no common zeros. Assume that (5)holds. Then there are two entire functions $H_1(z)$ and $H_2(z)$ satisfying

$$f_j(h(z))^2 g_j(h(z)) = f_j(k(z))^2 g_j(k(z)) e^{2H_j(z)} \qquad (j = 1, 2)$$

Hence, putting $\alpha_j(z) = f_j(k(z)) e^{H_j(z)}/f_j(h(z))$, we have

$$g_j(h(z)) = \alpha_j(z)^2 g_j(k(z)) \qquad (j = 1, 2) \tag{8}$$

which are the same form as the functional equation (6). Here we may assume, without loss of generality, that $g_1(z) \not\equiv$ const. or $g_2(z) \not\equiv$ const., and if $f(z)$ is transcendental, then so is $g_1(z)$ or $g_2(z)$, that is, it has an infinite number of simple zeros. Because we can choose a constant A such that $G_1(z) = f(z) - A$ has simple zeros and $G_2(z) = 1/(f(z) - A)$ has simple poles (an infinite number if $f(z)$ is transcendental, and (5) implies $G_j(h(z)) = G_j(k(z))$ $(j = 1, 2)$. Therefore results of analytic mappings are almost rewritten as answers to the above problem.

Before we state the results on the above problem, we shall introduce some notations. Let $\overline{N}^*(r, a, f)$ be the counting function of a-points with odd multiplicity of $f(z)$ being not counted multiple and $(\overline{N}^*(r, a, f))$ the order of $\overline{N}^*(r, a, f)$. If $\overline{N}^*(r, 0, f)\overline{N}^*(r, \infty, f) \not\equiv 0$, we set $\rho^*(f) = \min\{\rho(\overline{N}^*(r, 0, f)), \rho(\overline{N}^*(r, \infty, f))\}$. If $\overline{N}^*(r, 0, f)\overline{N}^*(r, \infty, f) \equiv 0$, let $\rho^*(f)$ be the order of the counting function being not identical zero between $\overline{N}^*(r, 0, f)$ and $\overline{N}^*(r, \infty, f)$. We remark that $g_1(z)$ in (7) is uniquely determined for $f(z)$ if $\rho(\overline{N}^*(r, 0, f)) < \infty$ and $g_2(z)$ in (7) is unique determined if $\rho(\overline{N}^*(r, \infty, f)) < \infty$. We need

LEMMA A (Goldstein [4]). *Let $f_p(z)$ be a non-constant rational function of order* p *and* $g(z)$ *a non-constant meromorphic function. Then*

$$T(r, f_p(g)) \sim pT(r, g) \quad (r \to \infty)$$

No Now from Theorem B and Lemma A we can deduce that

THEOREM G. *Let $F(z)$ and $f(z)$ be non-constant meromorphic functions. If $h(z)$ is a polynomial (a transcendental entire function) satisfying* (5), *then $k(z)$ in* (5) *is also a polynomial (a transcendental entire function).*

By virtue of Theorem F, the analogue of the result of the author [10] is not valid. The analogue of the result of Baker [1] is as follows:

THEOREM H. *If $F(z)$, $f(z)$ are non-constant meromorphic functions with $\rho^*(f) < \infty$, and $h(z)$, $k(z)$ are non-constant polynomials satisfying* (5), *then either* (i) *there exist a root of unity λ and a constant β such that $k(z) = \lambda h(z) + \beta$ or* (ii) *there exist a polynomial $r(z)$ and constant c, s, t such that $h(z) = r(z)^2 + s$, $k(z) = (r(z) + c)^2 + t$.*

This theorem follows from the result of Baker [1] if there is a transcendental entire function $g_j(z)$ ($j = 1$ or 2) in (7) such that $\rho^*(f) = \rho(g_j)$. If either $g_1(z)$ or $g_2(z)$, say $g_2(z)$, is a non-constant polynomial, then $\alpha_2(z)$ in (8) must be constant, and so it has modulus 1 from Theorem F. Hence Theorem H can be proved by the argument of Baker-Gross [2] in the proof of Theorem D.

All the cases mentioned in Theorem H do occur.

Case (i) *with* $\lambda = 1$. Let $h(z)$ be a polynomial, $k(z) = h(z) + \beta$, $f(z) = \tan(\pi z/\beta)$. Then we have $F(z) = \tan(\pi h(z)/\beta) = f(h(z)) = f(k(z))$.

Case (i) *with* $\lambda \neq 1$. Then λ is a primitive j-th root of unity for some $j > 1$. Let $g(z)$ be a moermorphic function, $\eta = \beta/(1 - \lambda)$ and $f(z) = g((z - \eta)^j)$. Let $h(z)$ be a polynomial and $k(z) = \lambda h(z) + \beta$. Then we have $k(z) - \eta = \lambda(h(z) - \eta)$. Hence we have $F(z) = g((h(z) - \eta)^j) = f(h(z)) = f(k(z))$.

Case (ii) *with* $s = t$. Let $g(z) = \cos(2\pi\sqrt{z}/c)$ and $f(z) = g(z-s)/(g(z-s) - a)$, where a is a non-zero constant. For a polynomial $r(z)$, we put $h(z) = r(z)^2 + s$ and $k(z) = (r(z) + c)^2 + s$. Then we have $F(z) = \cos(2\pi r(z)/c)/(\cos(2\pi r(z)/c) - a) = f(h(z)) = f(k(z))$.

Next we shall consider the case when $h(z)$ and $k(z)$ in (5) are transcendental. If $f(z)$ is a rational function, then Lemma A implied that

THEOREM I. *If* $f(z)$ *is a non-constant rational function, and* $h(z)$, $k(z)$ *are transcendental meromorphic functions satisfying* (5), *then* $h(z)$ *and* $k(z)$ *are of the same order, the same type, and the same class.*

When $f(z)$ in (5) is transcendental, we can deduce from Theorem 1 and Theorem 2, respectively, that

THEOREM 1'. *Let* $f(z)$ *be a transcendental meromorphic function with* $\rho^*(f) < \infty$. *Assume that* $g_1(z)$ *in* (7) *is of positive lower order and*

$$T(r, g_1) < c\, N(r, 0, g_1) = c\, \overline{N}^*(r, 0, f) \quad \textit{for all} \quad r \geqq r_0$$

if $\rho^*(f) = \rho(\overline{N}^*(r, 0, f))$ *and* $g_2(z)$ *in* (7) *is of positive lower order and*

$$T(r, g_2) < c\, N(r, 0, g_2) = c\, \overline{N}^*(r, \infty, f) \quad \textit{for all} \quad r \geqq r_0$$

if $\rho^*(f) = \rho(\overline{N}^*(r, \infty, g))$, *where* c *is a positive constant.*

If transcendental entire functions $h(z)$ *and* $k(z)$ *of finite order satisfy* (5), *then they are of the same order*

THEOREM 2'. *Let* $f(z)$ *be a transcendental meromorphic function satisfying the assumptions of* Theorem 1', *and* $h(z)$ *and* $k(z)$ *transcendental entire functions satisfying* (5). *If* $h(z)$ *is of finite order* λ_h, *then the lower order* μ_k *of* $k(z)$ *is not greater than* λ_h.

3. LEMMAS

In order to prove our theorems we need several lemmas. From the argument in Ozawa [13, p. 2] and Mutō [7, p. 457], we can deduce

LEMMA B. *Let* $G(z)$ *and* $g(z)$ *be two entire functions, each of which has an infinite number of simple zeros and no other zeros. Suppose that there are two entire functions* $f(z)$ *and* $h(z)$ *satisfying*

$$f(z)^2G(z) = g(h(z))$$

Then for an arbitrarily small positive number ε *there is a positive number* r_0 *such that*

$$(1 - \varepsilon)N(r, 0, g(h)) \leqq N(r, 0, G) \leqq N(r, 0, g(h)) \text{ for all } r \leqq r_0,\ r \notin E$$

where E *is a set of finite measure depending only on* $h(z)$ *if* $h(z)$ *is of infinite order, and where* E *is empty if* $h(z)$ *is of finite order.*

LEMMA C ([11]). *Let* $g(z)$ *and* $h(z)$ *be entire functions. If* $M(r, h) > ((2 + \varepsilon)/\varepsilon)|h(0)|$ *for any* $\varepsilon > 0$, *then we have*

$$T(r, g(h)) \leqq (1 + \varepsilon)\ T(M(r, h), g)$$

LEMMA D ([11]). *Let* $g(z)$ *be a transcendental entire function,* $h(z)$ *a transcendental entire function of finite order,* η *a constant satisfying* $0 < \eta < 1$, *and* α *a positive number. Then we have*

$$N(r, 0, g(h)) \geqq (\log \frac{1}{\eta}) \left(\frac{N(M((\eta r)^{1/(1+\alpha)}, h), 0, g)}{\log M((\eta r)^{1/(1+\alpha)}, h) - O(1)} - O(1) \right)$$

as $r \to \infty$ *through all values.*

From the proof of Lemma 1 in Clunie [3], we deduce the following (cf. [11, Lemma 2]):

LEMMA E. *Let* $k(z)$ *be a transcendental entire function,* A *a positive number, and* $\alpha(r)$ *and* $\beta(r)$ *unbounded, strictly increasing, continuous functions satisfying*

$$\alpha(r) \geqq r, \ \beta(r) \geqq r \quad \text{and} \quad \log\beta(\eta\alpha(r)) = o(T(\xi r, k)) \quad (r \to \infty)$$

where η *and* ξ *are constants satisfying* $\eta > 1$ *and* $0 < \xi < 1$. *Then there are a positive number* R_0 *and an increasing sequence* $\{r_\nu\}_{\nu=1}^{\infty}$ *with* $r_1 > R_0$ *and* $r_\nu \to \infty$ $(\nu \to \infty)$ *such that for* $\nu \geqq 1$ *and for all* r *in* $r_\nu \leqq r \leqq \alpha(r_\nu)$ *and all* w *satisfying* $\beta(R_0) \equiv R_1 \leqq |w| \leqq \beta(r)$ *we have*

$$n(r, w, k) > A$$

4. PROOF OF THEOREM 1

Suppose, to the contrary, that the order λ_1 of $h_1(z)$ is different from the order λ_2 of $h_2(z)$. We may assume that $\lambda_1 < \lambda_2$.

It follows from Theorem A that the following functional equations

$$f_1(z)^2G(z) = g(h_1(z)) \quad \text{and} \quad f_2(z)^2G(z) = g(h_2(z))$$

hold with suitable entire functions $f_1(z)$ and $f_2(z)$. Hence by virtue of Lemma B and Lemma C, we have for all $r \geqq r_0$

$$(1-\varepsilon)N(r, 0, g(h_2)) \leqq N(r, 0, g(h_1)) \leqq (1+\varepsilon)T(M(r, h_1), g) \qquad (4.1)$$

Since $g(z)$ is of finite order and positive lower order, there are two positive numbers μ and λ satisfying

$$r^\mu < T(r, g) < r^\lambda \quad \text{for all} \quad r \geqq r_0 \qquad (4.2)$$

We choose positive numbers α, σ, τ, ξ, and ρ satisfying $\lambda_1 < \sigma < \tau < \xi/(1+\alpha) < \xi < \lambda_2 < \rho < +\infty$. Then we have for all $r \geqq r_0$

$$M(r, h_1) < \exp r^\sigma \quad \text{and} \quad M(r, h_2) < \exp r^\rho \qquad (4.3)$$

and there is an unbounded, increasing sequence $\{t_\nu\}$ of positive numbers satisfying

$$M((\eta t_\nu)^{1/(1+\alpha)}, h_2) > \exp\{(\eta t_\nu)^{\xi/(1+\alpha)}\} > \exp(\eta^\tau t_\nu^{\ \tau}) \qquad (4.4)$$

where η is a constant satisfying $0 < \eta < 1$. Using Lemma D together with (3), (4.2), (4.3), and (4.4), we have, with a suitable positive constant A,

$$(1-\varepsilon)\ N\ (t_\nu,\ 0,\ g(h_2)) > At_\nu^{-\rho}T(M((\eta t_\nu)^{1/(1+\alpha)},\ h_2),\ g) \qquad (4.5)$$

$$> At_\nu^{-\rho}T(\exp(\eta^\tau t_\nu^\tau),\ g)\ )$$

$$> At_\nu^{-\rho}\exp(\mu\eta^\tau t_\nu^\tau)$$

On the other hand, it follows from (4.2) and (4.4) that

$$T(M(r,\ h_1),\ g) < \exp(\lambda r^\sigma)$$

and consequently from (4.1) and (4.5) that

$$At_\nu^{-\rho}\exp(\mu\eta^\tau t_\nu^\tau) < (1+\varepsilon)\exp(\lambda t_\nu^\sigma)$$

that is,

$$(A/(1+\varepsilon))t_\nu^{-\rho}\exp\{t_\nu^\sigma(\mu\eta^\tau t_\nu^{\tau-\sigma} - \lambda)\} < 1$$

It is untenable, since the term in the left hand side is unbounded as $t_\nu \to \infty$. Therefore $\lambda_1 = \lambda_2$, which is our desire. Q.E.D.

5. PROOF OF THEOREM 2

Suppose, to the contrary, that there is an analytic mapping ϕ of R into S such that the projection k(z) of ϕ is of lower order μ_k greater than λ_h. Then from Theorem A we have two functional equations

$$f(z)^2G(z) = g(h(z)) \quad \text{and} \quad F(z)^2G(z) = g(k(z))$$

with two suitable entire functions f(z) and F(z). It follows from Lemma B and Lemma C that

$$(1-\varepsilon)\ N\ (r,\ 0,\ g(k)) \leqq N(r,\ 0,\ g(h)) \leqq T(r,\ g(h)) + O(1) \qquad (5.1)$$

$$\leqq (1+\varepsilon)\ T\ (M(r,\ h),\ g) \qquad (r \to \infty,\ r \notin E)$$

Now choose numbers σ, τ, and μ satisfying $1 < \sigma$, $1 < \tau$, and $0 \leqq \lambda_h < \mu < \sigma\tau\mu < \mu_k \leqq +\infty$. Put $\alpha(r) = r^{\sigma\tau}$ and $\beta(r) = \exp r^\mu$. Then we have $\log \beta(\eta\alpha(r)) = \eta^\mu r^{\sigma\tau\mu} = o(T(\xi r,\ k)) \quad (r \to \infty)$. Hence we apply Lemma E to k(z) with $A = 4c\mu(1+\varepsilon)/(1-\varepsilon)$, $\alpha(r) = r^{\sigma\tau}$ and $\beta(r) = \exp r^\mu$. Let $\{r_\nu\}_{\nu=1}^\infty$ be a sequence satisfying the statement of Lemma E. Let $\{w_j\}$ be the zeros of g(z). Choose t_ν satisfying $r_\nu^\sigma \leqq t_\nu \leqq r_\nu^{\sigma\tau}$ and $t_\nu \notin E$. Then we have for large ν that

$$N(t_\nu, 0, g(k)) \geqq \int_{r_\nu}^{t_\nu} \frac{n(t, 0, g(k))}{t} dt$$

$$\geqq \int_{r_\nu}^{t_\nu} \frac{1}{t} \{ \sum_{R_1 \leqq |w_j| \leqq \exp t_\nu^\mu} n(t, w_j, k)\} \; dt$$

$$\geqq \frac{4c\mu(1+\varepsilon)}{1 - \varepsilon} \int_{r_\nu}^{t_\nu} \frac{n(\exp t^\mu, 0, g) - n(R_1, 0, g)}{t} dt$$

$$\geqq \frac{4c\mu(1+\varepsilon)}{1 - \varepsilon} \int_{r_\nu^\mu}^{t_\nu^\mu} \frac{n(s, 0, g)}{\mu \log s} \frac{ds}{s} - O(\log t\ )$$

$$\geqq (4c(1+\varepsilon)/(1-\varepsilon))\ t_\nu^{-\mu} \{N (\exp t_\nu^\mu, 0, g)$$

$$- N (\exp r_\nu^\mu, 0, g)\} - O(\log t_\nu)$$

Since $N(r, 0, t)$ is convex in $\log r$, we have for large ν that

$$\frac{N(\exp t_\nu^\mu, 0, g)}{\log \exp t_\nu^\mu} \geqq \frac{N(\exp r_\nu^\mu, 0, g)}{\log \exp r_\nu^\mu}, \text{ or}$$

$$N(\exp t_\nu^\mu, 0, g) \geqq r_\nu^{\mu(\sigma-1)} N(\exp r_\nu^\mu, 0, g) \geqq 2N(\exp r_\nu^\mu, 0, g)$$

Hence we have

$$N(t_\nu, 0, g(k)) \geqq (2c(1+\varepsilon)/(1-\varepsilon))t_\nu^{-\mu} N (\exp t_\nu^\mu, 0, g) - O(\log t_\nu)$$

$$\geqq (c(1+\varepsilon)/(1-\varepsilon))t_\nu^{-\mu} N (\exp t_\nu^\mu, 0, g)$$

and consequently, taking (3) into account,

$$(1 - \varepsilon)\ N\ (t_\nu, 0, g(k)) \geqq (1 + \varepsilon)t_\nu^{-\mu}\ T(\exp t_\nu^\mu, g)$$

Thus combining with (5.1) we obtain for large ν that

$$t_\nu^{-\mu} T(\exp t_\nu^\mu, g) \leqq T(M(t_\nu, h), g) \tag{5.2}$$

On the other hand, since $h(z)$ is of finite order λ_h and $g(z)$ is of finite order and of positive lower order, there are three positive numbers λ, γ, and ρ such that $\lambda_h < \lambda < \mu$ and

$$M(r, h) < \exp r^{\lambda} \quad \text{and} \quad r^{\gamma} < T(r, g) < r^{\rho}$$

hold for large r. Hence it follows from (5.2) that for large ν

$$t_{\nu}^{-\mu}(\exp t_{\nu}^{\mu})^{\gamma} \leqq (\exp t_{\nu}^{\lambda})^{\rho}$$

and consequently

$$t_{\nu}^{-\mu}\exp \{t_{\nu}^{\lambda}(\gamma t_{\nu}^{\mu-\lambda} - \rho)\} \leqq 1$$

It is untenable. Therefore there is no analytic mapping of R into S such that its projection is of lower order greater than λ_h. Q.E.D.

REFERENCES

1. I. N. Baker, Analytic mappings between two ultrahyperelliptic surfaces, *Aequationes Math. 14* (1976), 461-572.

2. I. N. Baker and F. Gross, On factorizing entire functions, *Proc. London Math. Soc. (3) 18* (1968), 69-76.

3. J. Clunie, The composition of entire and meromorphic functions, *Mathematical Essays Dedicated to A. J. Macintyre* (Ohio University Press, 1970), 75-92.

4. R. Goldstein, On deficient values of meromorphic functions satisfying a certain functional equation, *Aequationes Math. 5* (1970), 75-84.

5. W. K. Hayman, *Meromorphic Functions,* Clarendon Press, Oxford, 1964.

6. G. Hiromi and H. Mutō, On the existence of analytic mappings, I, *Kōdai Math. Sem. Rep. 19* (1967), 236-244.

7. H. Mutō, On the family of analytic mappings among ultrahyperelliptic surfaces, *Kōdai Math. Sem. Rep. 26* (1974/75), 454-458.

8. H. Mutō and K. Niino, A remark on analytic mappings between two ultrahyperelliptic surfaces, *Kōdai Math. Sem. Rep. 26* (1974/75), 103-107.

9. K. Niino, On the family of analytic mappings between two ultrahyperelliptic surfaces *Kōdai Math. Sem. Rep. 21* (1969), 182-190.

10. K. Niino, On the family of analytic mappings between two ultrahyperelliptic surfaces, II, *Kōdai Math. Sem. Rep. 21* (1969), 491-495.

11. K. Niino and N. Suita, Growth of a composite function of entire functions, *Kōdai Math. J. 3* (1980), No. 3, 374-379.

12. M. Ozawa, On complex analytic mappings, *Kōdai Math. Sem. Rep. 17* (1965), 93-102.

13. M. Ozawa, On the existence of analytic mappings, *Kōdai Math. Sem. Rep.* *17* (1965), 191-197.

14. M. Ozawa, On the existence of analytic mappings, II, *Kōdai, Math. Sem. Rep.* *18* (1966), 1-7.

EXTENDING NEVANLINNA'S RAMIFICATION RESULTS

Charles F. Osgood

Naval Research Laboratory
Washington, D. C.

INTRODUCTION

In the present paper, part of the ramification theory due to Nevanlinna is extended to the case in which small algebroid functions take the place of constant values. In particular, it is shown that no meromorphic function can be completely ramified at more than 4 "small" algebroid points.

SECTION I

THEOREM I. Suppose that f is a meromorphic function. Then there can exist at most 4 distinct algebroid functions α_j, each having the two properties $T(r,\alpha_j) = o(T(r,f))$ and every zero of $f-\alpha_j$ is ramified.

THEOREM II. If f is entire, the number of distinct α_j in Theorem I is at most 2.

Actually more is true, some of which will be explicitly noted in the proofs. The function f can also be a non-meromorphic algebroid function if the number of singular points is "very small." Additionally, a "very small" number of zeros of each $f-\alpha_j$ need not be ramified. Finally, it will become clear from the proof that the requirement that each $T(r,\alpha_j)$ be $o(T(r,f))$ is not quite necessary. Implicit in the proof below is that having each $T(r,\alpha_j)$ be eventually less than $\varepsilon T(r,f)$, for some $\varepsilon > o$ (which *could* even be explicitly calculated) suffices to ensure that Theorems I and II hold.

Nevanlinna's results corresponding to Theorems I and II can be proven *with ease* from his bounds on the sum of all of the ramifications (at constant values) and on the sum of the defects (at constant values). Continuing work begun in [1], I can obtain bounds analogous to those of Nevanlinna, but *weaker*, for both the sum of the ramifications and the sum of the defects at small meromorphic values. However, obtaining my Theorems I and II from my *weak* defect and ramification results is a rather close thing; I almost fail.

If it is hypothesized, additionally, that each zero of every $f-\alpha_j$ has multiplicity $\geq n > 2$, stronger bounds on the number of completely ramified algebroid "points" α_j may sometimes be shown, using the methods introduced here. As an example, we shall prove that if f is entire and $n \geq 3$, there can be at most one such α_j, α_1, and further α_1 must be meromorphic.

At a crucial place in the proofs I was helped considerably by a suggestion, due to Fred Gross, that I consider taking roots. Fred Gross and I are writing a paper which, in part, applies the present results to functional equations.

SECTION II

Motivated by Selberg's work in [2,3], we shall next define a counting function $\bar{\bar{n}}$. Suppose that $z = a$ is a branch point of an algebroid function g. If $g(z) \neq \infty$, near $z = a$,

$$g(z) = g(a) + \sum_{j=j_o}^{\infty} \gamma_j (z-a)^{j/q}$$

for some integers $j_o \geq 0$ and $q > 0$, where $\gamma_{j_o} \neq 0$. If $g(a) = \infty$,

$$g(z) = \sum_{j=j_o}^{\infty} \gamma_j (z-a)^{j/q}$$

where now $j_o < 0$. Let $\bar{\bar{n}}(r)$ equal the number of points $z = a$ having absolute value at most r such that $z = a$ is a branch point of g *and* $|j_o q^{-1}| < 1$. The points are to be counted once on each of the at most q sheets joined at $z = a$.

DEFINITIONS. Set

$$N(r,g) = k^{-1} \int_o^r \frac{\bar{\bar{n}}(t)-\bar{\bar{n}}(o)}{t}\, dt + \bar{\bar{n}}(o)k^{-1}\log r$$

If g is an algebroid function and $\bar{\bar{N}}(r,g) = o(T(r,g))$, we say that g is *nearly meromorphic*. If also $N(r,g) = o(T(r,g))$, we say that g is *nearly entire*.

Selberg in [2,3] defined functions $T(r,g)$, $m(r,g)$, and $N(r,g)$ for algebroid functions g. He defined a counting function $n(t)$ such that each pole is counted with its multiplicity on every sheet of the k-sheeted surface associated with g. He set

$$N(r,g) = k^{-1} \int_o^r \left(\frac{n(t)-n(0)}{t}\right) dt + k^{-1}\, n(o)\log r$$

and

$$m(r,g) = (2\pi k)^{-1} \int_o^{2\pi} \left(\sum_{j=1}^{k} \log^{+}|g_j(re^{i\phi})|\right)d\phi$$

where the g_j are the k different algebraic conjugates of g. Selberg defined the characteristic function of g by $T(r,g) = N(r,g) + m(r,g)$. He proved the first and second funcamental theorems for his T function: from what he proved, we know that

$$T(r,g) = T(r,(g-\alpha)^{-1}) + O(1) = T(r,g-\alpha) + O(1)$$

for all constants α and $m(r,g'/g) = o(T(r,g))$, except (possibly) on a set of finite measure.

LEMMA I. If f is nearly meromorphic, then

$$T(r,f') \leq 2T(r,f) + o(T(r,f))$$

except possibly on a set of finite measure.

Proof. Let k denote the degree of f. Since

$$m(r,f') \leq m(r,f) + m(r,f'/f) = m(r,f) + o(T(r,f))$$

except for a set of finite measure, all that is needed is to prove that

$$N(r,f') \leq 2N(r,f) + o(T(r,f))$$

Recall the previously defined functions $n(t) = n(t,f)$ and $\bar{\bar{n}}(t) = \bar{\bar{n}}(t,f)$. Clearly $n(t,f') \leq 2n(t,f) + (k-1)\bar{\bar{n}}(t,f)$. Thus $N(r,f') \leq 2N(r,f) + o(T(r,f))$. This proves Lemma I.

Next we shall prove:

LEMMA II. If $T(r,\alpha_j) = o(T(r,f))$, then $T(r,\alpha_j') = o(T(r,f))$.

Proof. Using the proof of Lemma I, $T(r,\alpha_j') = m(r,\alpha_j') + N(r,\alpha_j')$ $\leq m(r,\alpha_j) + o(T(r,f)) + N(r,\alpha_j') = o(T(r,f)) + N(r,\alpha_j')$ $\leq o(T(r,f)) + (2N(r,\alpha_j) + (k-1)\,\bar{\bar{N}}(r,\alpha_j)) = o(T(r,f))$.
This proves Lemma II.

DEFINITIONS. Let $n_\ell(t)$ denote the number of poles of f of order *larger than* ℓ having absolute value no larger than t, each counted (on every sheet where it appears) to the larger of zero and the multiplicity minus ℓ.

Let

$$N_\ell(r,f) = (2\pi k)^{-1} \int_0^r \left(\frac{n_\ell(t)-n_\ell(o)}{t}\right) dt + k^{-1}\, n_\ell(o)\log r$$

LEMMA III. If each of the algebroid functions α_1, α_2, and α_3 has a characteristic function which is $o(T(r,f))$, then $\sum_{j=1}^{3} (m(r,(f,\alpha_j)^{-1}) + N_1(r,(f-\alpha_j)^{-1}))$ $\leq (2+o(1))\,T(r,f)$, outside of (possibly) a set of finite measure.

Proof. We begin by constructing a nonzero generalized Riccati differential equation

$$P(y) = ay' + by^2 + cy + d = 0$$

such that

$$P(\alpha_1) \equiv P(\alpha_2) \equiv P(\alpha_3) \equiv 0$$

This set of requirements leads to a system of 3 linear homogeneous equations in the four unknown functions a, b, c, and d. Thus there exists a nontrivial solution (a, b, c, d) with each component being in the ring generated over the complex field by α_1, α_1', α_2, α_2', α_3, and α_3'. Using Lemma II, and obvious inequalities about the Selberg T function, we see that a, b, c, and d must have characteristic functions which are $o(T(r,f))$.

By an argument given in Corollary II of [1] (changed only to replace Nevanlinna's T function by Selberg's T function), it follows that f cannot be a solution of any (nontrivial) generalized Riccati equation

$$P(y) = 0$$

which has the 3 distinct "small" functions α_1, α_2, and α_3 as solutions.

We need to estimate T(r,P(f)) from above and below. If f, α_1, α_2, and α_3 were each meromorphic functions, Theorem 1 of [1] would apply (since also a, b, c, and d would be meromorphic) and one would have

$$\sum_{j-1}^{3} (N_1(r,(f-\alpha_j)^{-1}) + m(r,(f-\alpha_j)^{-1})) \leq (2+o(1))T(r,f)$$

outside of a set of finite measure, which is the result that we need.

In [4] (see Theorem IV), Theorem I of [1] was extended to handle the case of algebroud α_j if the coefficients (in the present case a, b, c, and d) are meromorphic. But our a, b, c, and d are not (necessarily) meromorphic. Thus we must see what difficulties, if any, exist in extending Theorem 1 of [1] to the case of an algebroid f *and* algebroid coefficients. *Difficulties do arise, in general, or the extension would have been given in* [4]. What is needed for there to be no change in the conclusion is exactly the hypothesis that f,..., $f^{(n-1)}$ are each nearly meromorphic, where n is the order of P(y). We need some notation: Let d_j be the "order of vanishing" of P(y) at $y^{(s)} = \alpha_j^{(s)}$ for s = 0, 1,...; i.e., each of the mixed partial derivatives of P(y) (with respect to derivatives of y) of order less than d_j vanishes at $y^{(s)} = \alpha_j^{(s)}$, for s = 0, 1,.... The numbers m_1 and d shall denote, respectively, the degree of P(y) in y and its derivatives, and the denomination of P(y). (The definition of the denomination is given in [1].) We can now conclude the proof of Lemma III with:

THEOREM III. Theorem I of [1] holds when each α_j is algebroid, and f is algebroid, if additionally, $f, f^{(1)}, \ldots, f^{(n-1)}$ are each nearly meromorphic. Further, a stronger inequality holds:

$$\sum_{j=1}^{k} (d_j m(r,(f-\alpha_j)^{-1}) + d_j\, N_n(r,(f-\alpha_j)^{-1}))$$

$$\leq T(r,P(f)) + o(T(r,f))$$

$$\leq d(N(r,f)) + m_1(m(r,f)) + o(T(r,f))$$

In the setting of Lemma III, each $d_j = 1$, and d = 2. Thus Lemma III follows from Theorem III.

We shall actually prove a sometimes stronger result, Theorem IV.

DEFINITIONS. For each α_j, consider the set E_j of all

$(\Pi_s(\frac{\partial}{\partial y^{(s)}})^{e_s})P(y)$ which are nonzero at $y^{(s)} = \alpha_j^{(s)}$, for $s = 0,1,\ldots.$

Let ϕ_j denote the minimum over E_j of the numbers

$$(\sum_{s=0}^{n} (n-s)e_s + \sum_{s=0}^{n} e_s)-d_j$$

For each positive integer ℓ let $\overline{n}_\ell(t)$ count, without multiplicity, the number of poles having multiplicity greater than ℓ in the disk of radius t. Set

$$\overline{N}_\ell(r,f) = (2\pi k)^{-1} \int_0^r (\overline{n}_\ell(t) - \overline{n}_\ell(o))\, t^{-1}\, dt + k^{-1}\, \overline{n}_\ell(o) \log r$$

THEOREM IV (A stronger version of Theorem III). Under the hypotheses of Theorem III,

$$\sum_{j=1}^{k} (d_j m(r,(f-\alpha_j)^{-1}) + d_j N_n(r,(f-\alpha_j)^{-1})) + \sum_{j=1}^{k} \phi_j \overline{N}_n(r,f-\alpha_j)^{-1})$$

$$\leq T(r,(P(f))^{-1}) + o(T(r,f))$$

$$= T(r,P(f)) + o(T(r,f))$$

$$\leq fN(r,f) = m_1(m(r,f)) + o(T(r,f))$$

Proof of Theorems III and IV. We shall follow the outline of the proof of Theorem I of [1] indicating only the changes needed. The proof of Theorem I of [1] consists of 5 parts. The fifth part need not concern us since it simply combines the results of the first four parts while utilizing the fact that the first fundamental theorem holds for (Selberg's) T function. Intuitively speaking, part (i) is still valid because of the hypothesis that $f, f', \ldots, f^{(n-1)}$ are each almost meromorphic, since this hypothesis implies that "almost all" of the poles of the functions $f', \ldots, f^{(n)}$ are of order ≥ 1.

To be more precise, define a counting function $n(t,f^{(n)})$ which counts (with multiplicities and on each sheet of the Riemann surface): the number of poles of $f^{(n)}$ that are not poles of f, plus the number of poles of f having order less than one. This n function gives rise to a corresponding N function which must be $o(T(r,f))$, because of the hypothesis that $f^{(n-1)}$ is nearly meromorphic. This causes (i) to go through by the same proof, i.e., "essentially all" of the poles of P(f) are poles of f and they occur to an order at most d times the order of the corresponding pole in f. Parts (ii) and (iii) go through with no changes except that in part (iii) we can use

$$\sum_{j=1}^{k} d_j N_n(r,(f-\alpha_j)^{-1}) + \sum_{j=1}^{k} \phi_j \overline{N}_n(r,(f-\alpha_j)^{-1})$$

as a lower bound for $N(r,(P(f))^{-1})$, replacing $\sum_{j=1}^{k} N_n(r,(f-\alpha_j)^{-1}) + o(T(r,f))$

which was used in the proof of Theorem I of [1]. (To see the new lower bound, expand P(f) in a multivariable power series in the "variables" $(f-\alpha_j)^{(\ell)}$ and estimate the number of zeros common to each term.) This new bound will give Theorem IV. Part (iv) goes through with the changes noted in [4] to prove the extension of Theorem I of [1] stated there. This proves Theorems III and IV.

SECTION III

Proof of Theorem I. Assume without loss of generality that exactly five functions $\alpha_j(z)$ exist. By the linear fractional transformation sending y into $(y-\alpha_1)(y-\alpha_2)^{-1}$, we may take f into a nearly meromorphic function f_1 and the "small" algebroid functions into o, ∞, β_1, β_2, and β_3, respectively; the β_1, β_2, and β_3 are clearly "small" algebroid functions. We must check that f_1 is nearly meromorphic. Write f_1 as $1 + (\alpha_2-\alpha_1)(f-\alpha_2)^{-1}$; those branch points of f_1 which do not come from f can be safely ignored in what follows. We may as well restrict ourselves to looking at branch points z_o of f where α_1, α_2, and $\alpha_1-\alpha_2$ have no zeros or poles. Notice that if $f-\alpha_2$ does not either vanish or equal infinity at $z = z_o$ and if $j_o/q \geq 1$ there (for f), then $j_o/q \geq 1$ for f_1 also. If $f-\alpha_2$ equals zero and $j_o/q \geq 1$ (for f) then, since $\alpha_1-\alpha_2$ does not equal o or ∞ at $z = z_o$, $j_o/q \leq -1$ for f_1. If $f = \infty$ at $z = z_o$ and $j_o/q \leq -1$ (for f) then, since $\alpha_1-\alpha_2$ does not equal zero or ∞ at $z = z_o$, $j_o/q \geq 1$ for f_1. It follows that f_1 is nearly meromorphic.

Because of the ramification hypothesis, it is easy to see that $\sqrt{f_1}$, $\pm\sqrt{\beta_1}$, $\pm\sqrt{\beta_2}$, $\pm\sqrt{\beta_3}$ form a set consisting of one nearly meromorphic function and six algebroid functions. We wish to see that if we define six functions n(t) to count, on all sheets of the appropriate Riemann surface, the number of *unramified* zeros of $\sqrt{f_1}$ minus any one of the six small algebroid functions, the corresponding N functions will each be o(T(r,f)). If say $\sqrt{f_1} - \sqrt{\beta_1}$

vanishes at $z = z_o$, then either the zero is at least double or β_1 vanishes at $z = z_o$. (The zeros of $(\sqrt{f_1} - \sqrt{\beta_1})$ are all ramified at $z = z_o$. Either the zero, with multiplicity, belongs completely to one factor, or β_1 also vanishes at $z = z_o$.) Since $N(r,\beta_1) = o(T(r,f)$, we need only contemplate the first possibility.

We are now in a position essentially the same as contemplated in our original hypotheses except for two changes: one change is minor and the other will turn out to be crucial. The unimportant change is that the ramification at our small functions is no longer complete but instead is merely "essentially complete," where the meaning is the obvious one. The content of this change is that we need only have demanded "essentially complete" ramification for $\alpha_1,\ldots,\alpha_5$ in our original hypotheses. The important change is that we now have six small algebroid functions which, regarded as values, are "essentially completely" ramified. Repetition of the above argument once more will give 2(6-2) = 8 small "essentially completely" ramified algebroid functions. We shall demonstrate a contradiction assuming 7 such functions. (This seems a good place to point out how the argument could be varied to handle a nearly entire function f and three completely ramified small algebroid values. We can argue that $\sqrt{f-\alpha_{\cdot}}$ is nearly entire. This gives us 4 "essentially completely" ramified small algebroid functions. Two more repetitions of the above argument gives 10 such functions.)

We wish to construct a differential equation of the form

$$P(y) = a(y)(y')^2 + b(y)y' + c(y) = 0 \tag{1}$$

where a(y), b(y), and c(y) are polynomials having degrees $\leq$ 1, 3, and 5, respectively. This means that 2+4+6-1 = 11 linear homogeneous conditions can be imposed upon the coefficients of a, b, and c. We require that (1) be satisfied by $\alpha_1,\ldots,\alpha_7$ and that

$$2a(y)y' + b(y) = 0 \tag{2}$$

be satisfied by $\alpha_1,\ldots,\alpha_4$ where, using Lemma III, α_5, α_6, and α_7 are chosen so that on a common set of r values having infinite measure

$$N(r,(f-\alpha_j)^{-1}) - N_1(r,(f-\alpha_j)^{-1}) > o(T(r,f))$$

for j = 5, 6, and 7. The coefficient functions in (2) must have characteristic functions which are o(T(r,f)). Suppose first that $P(f) \not\equiv o$. Then, using Theorem IV we have, since $d_j \geq 1$ if j = 1,2,...7 and $\phi_j \geq 1$ if j = 1, 2, 3, or 4, $d \leq 5$, and $m_1 \leq 5$,

$$\sum_{j=1}^{7} m(r,(f-\alpha_j)^{-1}) + \sum_{j=1}^{7} N_1(r,(f-\alpha_j)^{-1})$$

$$+ \sum_{j=1}^{4} \overline{N}_1(r,(f-\alpha_j)^{-1})$$

$$\leq T(r,P(f)) + o(T(r,f))$$

$$\leq 5N(r,f) + 5m(r,f) + o(T(r,f))$$

$$= (5+o(1))\ T(r,f)$$

Since each of the $f-\alpha_j$ have "essentially completely" ramified zeros (on every sheet of the Riemann surface corresponding to $f-\alpha_j$), it follows that

$$\sum_{j=1}^{7} m(r,(f-\alpha_j)^{-1}) + \sum_{j=1}^{7} N_1(r,(f-\alpha_j)^{-1}) + \sum_{j=1}^{4} \overline{N}_1(r,(f-\alpha_j)^{-1}) + o(T(r,f))$$

$$> (4+3/2)\ T(r,f) = 5\tfrac{1}{2}\ T(r,f)$$

This gives a contradiction. Thus either $P(f) \equiv 0$ or Theorem I is true. Suppose $P(f) \equiv 0$. For each j, if $\partial P/\partial y'(\alpha_j) \not\equiv 0$, $T(r,\partial P/\partial y'(\alpha_j))$. Therefore if $\partial P/\partial y'(\alpha_j) \not\equiv 0$, it follows from the Picard Existence and Uniqueness Theorem for differential equations that at all but (at most) $o(T(r,f))$ points with absolute value $\leq r$ where both $f-\alpha_j = 0$ and $(f-\alpha_j)' = 0$, $\partial P/\partial y'(\alpha_j)$ must vanish. (The exceptional at most $o(T(r,f))$ points are the zeros or poles of $\partial P/\partial y'(\alpha_j)$ and the poles of α_j.) However, using Lemma III, we have chosen α_5, α_6, and α_7 such that for j = 5, 6, and 7, $N(r,(f-\alpha_j)^{-1}) - N_1(r,(f-\alpha_j)^{-1}) > o(T(r,f))$ on a common set of infinite measure.

Thus we would have a contradiction unless $\partial P/\partial y'(\alpha_j) \equiv 0$, for j = 5, 6, and 7.

Now consider the discriminant of $P(y)$ regarded as a polynomial in y'. The discriminant of P has degree at most 6. Since all seven of the α_j are double zeros of $P(y)$, this implies that the discriminant must vanish identically. Under this condition, if $a(y) \equiv 0$, then $b(y) \equiv 0$ and also $c(y) \equiv 0$, contrary to assumption.

Therefore $a(y) \not\equiv 0$ and

$$P(y) = a(y)(y'-b(y)/2a(y))^2$$

It follows that

$$a(y)y' - b(y) = 0$$

for $y = f, \alpha_1,\ldots,\alpha_7$. We shall again apply the Picard Existence and Uniqueness Theorem for differential equations. Suppose for some complex number z_o, $f(z_o) = \alpha_j(z_o) \neq \infty$. Then, by Picard, since $f(z) \not\equiv \alpha_j$, either:

(i) $a(\alpha_j,z)$ is zero or infinite at $z = z_o$

or

(ii) $b(\alpha_j,z)$ is infinite at $z = z_o$.

If $a(\alpha_j,z) \not\equiv 0$, the points of absolute value $\leq r$ satisfying condition (i), condition (ii), or the condition

(iii) $\alpha_j(z_o) = \infty$

are fewer than

$$(N(r,a(\alpha_j,z)) + N(r,(a(\alpha_j,z))^{-1})) + (N(r,b(\alpha_j,z)) + N(r,(b(\alpha_j,z)^{-1})) +$$

$$(N(r,\alpha_j) = o(T(r,f))$$

From Lemma III we know that there exist two j's such that $N(r,(f-\alpha_j)^{-1}) - N_1(r,(f-\alpha_j)^{-1}) > o(T(r,f))$, on a common set of infinite measure. Recall that $a(y,z)$ is linear in y, so $a(\alpha_j,z) \equiv 0$ for at most one α_j. Therefore there exists a point z_o and a positive integer $j_o \leq 7$ such that

$$a(\alpha_{j_o},z) \not\equiv 0$$

$$(f-\alpha_{j_o})(z_o) = 0$$

$$\alpha_j(z_o) \neq \infty$$

$a(\alpha_{j_o},z)$ does not equal 0 or ∞ at $z = z_o$, and $b(\alpha_{j_o},z)$ does not equal ∞ at $z = z_o$. By Picard, $f \equiv \alpha_{j_o}$. This contradiction shows $P(f) \not\equiv 0$. Theorem I now follows.

Because of the remarks made in the proof of Theorem I, Theorem II follows also.

The example at the end of Section I can now be dealt with swiftly: If the statement in our example is false, $\sqrt[3]{f-\alpha_1}$ is nearly entire and is, also, nearly completely ramified at each of the three cube roots of $\alpha_2 - \alpha_1$, contradiction. Thus only one such α_j, α_1, can exist. If α_1 were non-meromorphic, any (algebraic) conjugate of α_1, say $\tilde{\alpha} \neq \alpha_1$, would satisfy the hypotheses required of α_2, contradiction.

REFERENCES

1. C. F. Osgood, A number theoretic - differential equations approach to generalizing Nevanlinna theory, *Indian Journal of Mathematics, 23, No. 1* (1980).

2. H. L. Selberg, über eine eigenschaft der logarithmischen ableitung einer meromorpher oder algebroiden function endlicher Ordnung, Avhandlinger Utgitt AV Det Norske Videnskaps-Akademic/Oslo I, *Matem Naturv. Klasse 14* (1929).

3. H. L. Selberg, Algebroide funktioner, *Ibid, 8* (1934).

4. F. Gross and C. F. Osgood, On the functional equation $f^n + g^n = h^n$ and a new approach to a class of more general functional equations, *Indian Journal of Mathematics, 23, No. 1* (1981).

SOME GROWTH RELATIONSHIPS ON FACTORS OF TWO COMPOSITE ENTIRE FUNCTIONS

Kiyoshi Niino

Faculty of Technology
Kanazawa University
Japan

Chung-Chun Yang

Naval Research Laboratory
Washington, D.C.

1. INTRODUCTION

A transcendental entire function F(z) is called composite iff F(z)=f(g) for some nonlinear entire functions (factors) f,g.In [2.p.257] Gross conjectured that every transcendental composite entire function must have infinitely many fix-points. The conjecture has not been completely settled yet. However, if some composite entire function F has only finitely many fix-points, then one has $F(z)-z \equiv f(g)-z \equiv p(z)e^{h(z)}$, where p is a polynomial and h is an entire function. It follows that

$$\log M(r,f(g)) \sim \log M(r,e^{h}) \text{ as } r \to \infty \quad . \tag{1}$$

Thus it is important to investigate the possible growth relationships between factors f and h as well as between factors g and h, whenever f, g, and h satisfy condition (1). In this connection, we have obtained a number of interesting related results in this note as follows.

THEOREM 1. Let f and g be entire functions which are of positive lower order and of finite order. If entire functions h and g satisfy

$$0 < \liminf_{r\to\infty} \frac{T(r,f(h))}{T(r,g(k))} \leq \limsup_{r\to\infty} \frac{T(r,f(h))}{T(r,g(k))} < +\infty \tag{2}$$

then ρ_h(order of h) = ρ_g(order of g).

THEOREM 2. Let h, f, g be entire functions with μ_f (lower order of f) $\leq \rho_f$ (order of f) $\leq +\infty$ and

$$\log M(r, e^h) \sim \log M(f(g)) \text{ as } r \to \infty \quad . \tag{3}$$

If $\rho_f < 1$, then

$$\liminf_{r\to\infty} \frac{M(r,h)}{M(r,g)} = 0 \text{ and } \liminf_{r\to\infty} \frac{\log M(r,h)}{\log M(r,g)} \leq \rho_f \quad . \tag{4}$$

if $\rho_f \geq 1$, then

$$\liminf_{r\to\infty} \frac{M(r,g)}{M(r,h)} = 0 \text{ and } \limsup_{r\to\infty} \frac{\log M(r,g)}{\log M(r,h)} \geq \mu_f \quad . \tag{5}$$

The terms M(r,) and T(r,) denote as usual, the maximum modulus and Nevanlinna characteristic respectively. In the sequel, all the functions that we shall deal with are transcendental entire. Also it is assumed that the reader is familiar with some well-known results on the relationships between M(r,f) and T(r,f) as well as the growth rate of composite functions (see e.g., [3, p. 18 and p. 50]).

2. PROOFS OF THEOREM 1 AND THEOREM 2

2.1. Proof of Theorem 1. According to the hypotheses on the orders and lower orders of f and g, we have, for $r > r_1$,

$$r^{\alpha} < \log M(r,f) < r^{\beta}, \text{ and } r^{\gamma} < \log M(r,g) < r\,\delta ,$$

where α, β, γ, and δ are constants satisfying $\alpha < \mu_f$, $\beta > \rho_f$, $\gamma < \mu_g$, and $\delta > \rho_g$. Suppose that $\rho_h > \rho_k$, then $\exists$ constants Σ and η with $\rho_h > \Sigma > \eta > \rho_k$ such that $r > r_1$, $\log M(r,k) < r^{\eta}$ and, for a sequence $\{r_n\}$, $\log M(r_n,h) \geq r_n$. Now for any $r > o$

$$T(r,f(g)) \geq \tfrac{1}{3}\log M(\tfrac{r}{2},f(h)) \geq \tfrac{1}{3}\log M(\tfrac{1}{8}M(\tfrac{1}{2}\,\tfrac{r}{2},h)+o(1),f) = \tfrac{1}{3}\,\{\tfrac{1}{8}M(\tfrac{r}{4},h)+o(1)\}^{\alpha}.$$

It follows that for $r > r_2$,

$$T(r_n,f(h)) \geq \frac{1}{3}\left(\frac{1}{9}\right)^{\alpha} M\left(\frac{r_n}{4}, h\right)^{\alpha} \geq \frac{1}{3}\left(\frac{1}{9}\right)^{\alpha}\left[\exp\left(\frac{r_n}{4}\right)^{\xi}\right]^{\alpha}$$

$$= \frac{1}{3}\left(\frac{1}{9}\right)^{\alpha}\exp\left(\frac{\alpha r_n^{\xi}}{4^{\xi}}\right) = A\exp B_n \tag{6}$$

where $A = \frac{1}{3}\left(\frac{1}{9}\right)^{\alpha}$ and $B_n = \frac{\alpha}{4^{\xi}} r_n^{\xi}$.

On the other hand,

$$T(r,g(k)) \leq \log M(M(r,k),g) \leq \{M(r,k)\}^{\delta} \leq (e^{r^{\eta}})^{\delta} = e^{\delta r^{\eta}} . \tag{7}$$

Since $\eta-\xi<0$, (6) and (7) yields

$$\frac{T(r_n,f(h))}{T(r_n,g(k))} \geq \frac{Ae^{B_n}}{e^{\delta r_n^{\eta}}} = A\exp\{B_n\left(1 - \frac{\delta}{B_n} r_n^{\eta}\right)\} \tag{8}$$

which tends to ∞ as $n \to \infty$. This is a contradiction. Hence we must have $\rho_h \leq \rho_k$. Similarly, one can prove $\rho_k \leq \rho_h$. Thus $\rho_k = \rho_h$ as asserted.

2.2 Proof of Theorem 2. We proceed to prove inequality (4). Set $\rho_f=d<1$. Then $\forall\varepsilon>0$ satisfying $d+\varepsilon<1$ we have, for $r > r_0$,

$$\log M(r,f) < r^{d+\varepsilon} .$$

Hence, for $r > r_1$,

$$\log M(r,f(g)) \leq M(r,g)^{d+\varepsilon} . \tag{9}$$

Now from a result of Clunie [1, p. 76(1.3)] and the fact that $\log M(r,e^z)=r$, there exists a sequence $\{r_n\}$ such that

$$\log M(r_n,e^h) \geq \log M((1+o(1))M(r_n,h),e^z) = (1+o(1))M(r_n,h) . \tag{10}$$

Hence

$$1 + \varepsilon \geq \frac{\log M(r_n,e^h)}{\log M(r_n,f(g))} \geq \frac{(1+o(1))M(r_n,h)}{M(r_n,g)^{d+\varepsilon}}$$

$$= \frac{M(r_n,h)}{M(r_n,g)} (1+o(1))M(r_n,g)^{1-d-\varepsilon} . \tag{11}$$

Since $M(r_n,g)^{1-d-\varepsilon} \to \infty$ as $n \to \infty$, it follows that

$$\lim_{r\to\infty} \frac{M(r_n,h)}{M(r_n,g)} = 0$$

and hence

$$\liminf_{r\to\infty} \frac{M(r,h)}{M(r,g)} = 0 \quad .$$

Clearly, by arguing as above, one also can derive

$$1 + \varepsilon \geq \frac{\log\log M(r_n,e^h)}{\log\log M(r_n,f(g))} \geq \frac{\log M(r_n,h)+o(1)}{(d+\varepsilon)\log M(r_n,g)} \tag{12}$$

Hence

$$\liminf_{r \to \infty} \frac{\log M(r,h)}{\log M(r,g)} \leq d \quad . \tag{13}$$

Inequality (4) is thus proved. Now it is clear that if one replaces $d+\varepsilon$ by $d-\varepsilon$, and reverses the sign of the inequality (9) (for a sequence $\{r_n\}$), then all the signs in inequality (11) will be reversed with $1+\varepsilon$ being replaced by $1-\varepsilon$. These observations lead to inequality (5). This also concludes the proof.

3. CONCLUDING REMARKS. (i). If condition (3) is substituted by a more general one:

$$0 < \liminf_{r \to \infty} \frac{\log M(r,e^h)}{\log M(r,f(g))} \leq \limsup_{r \to \infty} \frac{\log M(r,e^h)}{\log M(r,f(g))}$$

by employing the above arguments, one will get similar conclusions with modified numeric upper and lower bounds in inequalities (4), (5).

(ii). If condition (3) in Theorem 2 is relaxed to

$$0 < \liminf_{r \to \infty} \frac{\log M(r,e^h)}{\log M(r,f(g))} \leq \limsup_{r \to \infty} \frac{\log M(r,e^h)}{\log M(r,f(g))} < +\infty .$$

then the same argument can be carried over in this case with suitable modifications of the upper and lower bounds in inequalities (4), (5), (11), and (12).

REFERENCES

1. J. Clunie, The composition of entire and meromorphic functions, Math, Essays dedicated to A. J. Macintyre, Ohio University Press (1970), 75-92.

2. F. Gross, Factorization of meromorphic functions, U.S. Government Printing Office, Washington, D.C., 1972.

3. W. K. Hayman, Meromorphic functions, Oxford Press, 1964.

ON A CHARACTERIZATION OF THE EXPONENTIAL FUNCTION AND THE COSINE FUNCTION BY FACTORIZATION, IV

Mitsuru Ozawa

Department of Mathematics
Tokyo Institute of Technology
Oh-o kayama, Meguro-Ku, Tokyo
Japan

1. INTRODUCTION

It is well-known that e^z and cos z admit the factorizations $e^z = w^n(e^{z/n})$ and $\cos z = P_n(\cos z/n)$, respectively, with a suitable polynomial P_n of degree n for every n. This means that e^z and cos z occupy a quite special situation in the factorization theory. In our earlier papers ([2] [3]), we discussed the inverse problem of the above fact and proved the following theorems.

THEOREM A. Let F(z) be an entire function, which admits the following factorization

$$F(z) = P_m(f_m(z)) \qquad (*)$$

with a polynomial P_m of degree m and an entire function f_m for all integers of the form $m = 2^j$ and for m = 3. Then

$$F(z) = A \cos \sqrt{H(Z)} + B$$

unless

$$F(z) = A\, e^{H(z)} + B$$

Here H is a non-constant entire function and A, B are constants, $A \neq 0$.

THEOREM B. Suppose that (*) holds for $m = 3^j$ ($j = 1,2,\ldots$) and for $m = 2,4$. Then the same conclusion as in Theorem A holds.

Let $\{n_j\}$ be any infinite sequence of positive integers such that n_j is a true multiple of n_{j-1} for every j. As already remarked in [2], (*) only for $\{n_j\}$ does not imply the result as in Theorem A. However, it is very plausible to conjecture that (*) for $\{n_j\}$ and for q with $(q,n_j) = 1$ for all j implies the same result as in Theorem A.

In this paper we shall prove the following

THEOREM 1. Suppose that (*) holds for $m = 2^j$ and for $m = 5$, where j runs over all natural numbers. Then the same conclusion as in Theorem A holds.

The method in this paper suggests the following result: (*) for $m = 2^j$ and for $m = 2p+1$ implies the result as in Theorem A.

In order to prove Theorem 1, we make use of the following Lemmas repeatedly.

LEMMA 1. Let f(z) be an entire function. Then

$$(q-1)m(r,f) < \sum_{1}^{q} \overline{N}(r,a_k,f) + S(r)$$

where $a_k \neq \infty$ and $S(r) = O(\log rm(r,f))$ except for a set of finite measure.

LEMMA 2. Let f(z) be an entire function. Then

$$\sum_{a\neq\infty} \Theta(a) \leqq 1$$

where

$$1 - \Theta(a) = \overline{\lim_{r\to\infty}} \frac{\overline{N}(r,a,f)}{m(r,f)}$$

Let k(a) be the least order of almost all a-points of f(z). Then

$$\sum_{a\neq\infty} \left(1 - \frac{1}{k(a)}\right) \leqq 1$$

The following is the famous Picard theorem on uniformization of algebraic curves. See [4].

LEMMA 3. Let R be a closed Riemann surface of genus greater than one. Then it is impossible to uniformize R by any pair of meromorphic functions on the plane.

We further make use of the Riemann-Hurwitz relation in order to compute the genus of a Riemann surface.

2. PROOF OF THEOREM 1

The first step. Suppose $F(z)-b = A_2(f_2(z)-w_0)^2$ has only a finite number of zeros. Then $f(z)-b = Q(z)^2 e^{L(z)}$ with a polynomial Q and an entire function L = const. Further

$$F(z)-b = A_p \prod_1^p (f_p(z)-\alpha_j), \qquad p = 2^n$$

implies that all α_i must coincide with each other. Hence

$$F(z)-b = A_p(f_p(z) - \alpha_1)^p \quad \text{and} \quad f_p(z)-\alpha_1 = Q_p(z)e^{L_p(z)}$$

Therefore 2 deg Q = p deg Q_p holds for every n. Thus deg Q = deg Q^p = 0. Hence

$$F(z) = b+Ae^{L(z)}$$

with constants b, A. This is a part of our final result.

The second step. From now on we may assume that F-b has infinitely many zeros of even integral order. We consider the following case:

$$\begin{aligned} F-b &= A_2(f_2-w_0)^2 \\ &= A_4 (f_4 - w_1)(f_4 - w_2)(f_4 - w_3)^2 \\ &= A_5(f_5-d_1)(f_5-d_2)^2(f_5-d_3)^2 \end{aligned}$$

Here w_1, w_2, w_3 are different with each other and so are d_1, d_2, d_3. In this case we may put $f_4-w_1 = g^2$, $f_4-w_2 = h^2$, $f_5-d_1 = Y^2$. Then

$$g-h = \sqrt{w_2-w_1}\, e^{-L}, \qquad g+h = \sqrt{w_2-w_1}\, e^{L}$$

with an entire function L. Thus

$$g = 1/2 \sqrt{w_2-w_1}\, (e^L+e^{-L})$$

$$h = 1/2 \sqrt{w_2-w_1}\, (e^L-e^{-L})$$

This gives

$$A_2^{\frac{1}{2}}(f_2-w_o) = A_4^{\frac{1}{2}} \frac{(w_2-w_1)^2}{16} (e^{2L}-e^{-2L})(E^{2L}+d+e^{-2L})$$

$$= A_5^{\frac{1}{2}} Y(Y^2+d_1-d_2)(Y^2+d_1-d_3)$$

where

$$d = \frac{2(w_1+w_2-2w_3)}{w_2-w_1} \neq \pm 2$$

Let us put $X = e^{2L}$. We remark that almost all zeros of $X - a$ $(a \neq 0)$ are simple, exactly say

$$N(r,a,X) - \overline{N}(r,a,X) = o(m(r,X))$$

$$N(r,a,X) \sim m(r,X)$$

except for a set of finite measure. Next we wish to find out a constant D such that

$$Y^5+AY^3+BY+D = (Y-\beta_1)(Y-\beta_2)(Y-\beta_3)(Y-\beta_4)^2, \quad \beta_j \neq 0$$

Then

$$\beta_4^{\ 2} = 1/10\ (-3A \pm \sqrt{9A^2-20B})$$

$$D = 1/50\ (-6A^2+40B \pm 2A\sqrt{9A^2-20B})\beta_4$$

Thus, if $9A^2 \neq 20B$, there appear four values of D, all of which are different. We put them by D_k, $k = 1,2,3,4$. Then

$$Y^5+AY^3+BY+D_k = (Y-\beta_{k1})(Y-\beta_{k2})(Y-\beta_{k3})(Y-\beta_{k4})^2$$

Evidently β_{k4} $(k = 1,2,3,4)$ are multiple points of Y^5+AY^3+BY. Correspondingly we have

$$1/X^2(X +dX +\alpha_k X -dX-1) = 1/X^2(X-\alpha_{k1})(X-\alpha_{k2})(X-\alpha_{k3})^2$$

for

$$D_k = \alpha_k(A_4/A_5)^{\frac{1}{2}}(w_2-w_1)^2/16$$

Assume temporarily that α_{kj} $(k = 1,2,3,4;\ j = 1,2)$ are all different. Let R be the Riemann surface (X,Y) defined by

$$Y^5+AY^3+BY = C(X^2+dX-d/X-1/X^2), \quad C = D_k/\alpha_k$$

All the α_{kj} are branch points of order 2, $X = \infty$ is also a branch point of order 5, and $X = 0$ is an ordinary point. There is no other singular point of R. Hence by the well-known Riemann-Hurwitz relation

the genus of R = 2

Thus R is not uniformizable by any pair of meromorphic functions on the plane. However, X and Y are entire functions. Thus we have arrived at a contradiction.

Assume that $\alpha_{k1} = \alpha_{k2}$ for some k. Then by an easy calculation, this occurs only for $d = 0$ and $\alpha_k^2 = -4$. This case will be treated later.

We now consider the values of f_2 corresponding to $X = \alpha_{kj}$ and $X = \alpha_{k'n}$ $\alpha_{k'n}$ $(k' \neq k)$. Then these are different. Hence $\alpha_{kj} \neq \alpha_{k'n}$ for $k \neq k'$. Thus all the α_{kj} $k = 1,2,3,4$; $j = 1,2$ are different if $d \neq 0$.

Assume that $9A^2 = 20B$. Then there are only two values of D_k, for which

$$Y^5+AY^3+BY+D_k = (Y-\beta_{k1})(Y-\beta_{k2})(Y-\beta_{k3})(Y-\beta_{k4})^2, \quad k = 1,2$$

In this case there are either four values of α_k or only two values of α_k, for which

$$X^4+dX^3+\alpha_k X^2-dX-1 = (X-\alpha_{k1})(X-\alpha_{k2})(X-\alpha_{k3})^2, \quad \alpha_k \neq 0$$

since, if α_k is a required one, then so is $-\alpha_k$. If the former case occurs, then

$$\begin{aligned} Y^5+AY^3+BY+D_3 &= (Y-\beta_{31})(Y-\beta_{32})(Y-\beta_{33})(Y-\beta_{34})(Y-\beta_{35}) \\ &= C(X-\alpha_{31})(X-\alpha_{32})(X-\alpha_{33})^2/X^2 \end{aligned}$$

where $\beta_{3i} \neq \beta_{3j}$ for $i \neq j$. This gives

$$\sum_1^3 \overline{N}(r,\alpha_{3j},X) = \sum_1^5 \overline{N}(r,\beta_{3j},Y)$$

Hence

$$\frac{3}{4}5\ m(r,Y) \sim 3\ m(r,X) \geqq 4\ m(r,Y)(1+o(1))$$

which is clearly untenable. Therefore there are only two values of α_k. In this case it is easy to prove $d = \pm 2$, which had been excluded.

If $d = 0$, then α_k^2 must be -4 and

$$C(X^4 \pm 2i\, X^2 - 1)/X^2 = C(X^2 \pm i)^2 \,/\, X^2$$
$$= (Y - \beta_{k1})(Y - \beta_{k2})^2(Y - \beta_{k3})^2$$

Now we return back to the original equation

$$F\text{-}b = A_2(f_2 - w_o)^2$$
$$= A_4(f_4 - w_1)(f_4 - w_2)(f_4 - w_3)^2$$
$$= A^*(e^{4L} - e^{-4L})^2$$
$$= 2A^*(\cos K - 1), \qquad 8L = Ki, \qquad A^* = A_4(w_2 - w_1)^4 / \, 4^4$$

This is a part of our final result.

Still we need to discuss the following case:

$$C(X - \alpha_{k1})(X - \alpha_{k2})^3 \,/\, X^2$$
$$= (Y - \beta_{k1})(Y - \beta_{k2})(Y - \beta_{k3})^3$$

There appear four values of α_k and D_k. Further α_{11}, α_{21}, α_{31}, α_{41} are different. By the Riemann-Hurwitz relation the genus of R, defined by (X,Y), is equal to 2 and hence Lemma 3 is applicable. We have again a contradiction.

We consider the following case:

$$F - b = A_2(f_2 - w_o)^2 = A_4(f_4 - w_1)(f_4 - w_2)(f_4 - w_3)^2$$
$$= A_5\,(f_5 - d_1)(f_5 - d_2)^4 \quad \text{or} \quad A_5\,(f_5 - d_1)^3(f_5 - d_2)^2$$

As in the above case

$$F - b = A_4\frac{(w_2 - w_1)^4}{16^2}(e^L + e^{-L})^2(e^L - e^{-L})^2(e^{2L} + d + e^{-2L})^2$$

Hence

$$\bar{N}\,(r, w_3, f_4) \sim m(\,r, f_4)$$
$$\bar{N}\,(r, w_1, f_4) \sim N(\,r, w_1, f_4)/2 \sim m(\,r, f_4)/2$$
$$\bar{N}\,(r, w_2, f_4) \sim N(\,r, w_2, f_4)/2 \sim m(\,r, f_4)/2$$

Thus

$$\sum_1^3 \bar{N}(r, w_j, f_4) = \bar{N}(r, d_1, f_5) + \bar{N}(r, d_2, f_5)$$

implies that

$2m(r, f_4) \leqq 2m(r, f_5)(1 + o(1)) \sim 8m(r, f_5)/5$, which is clearly untenable.

Next we consider the case $d = \pm 2$, that is,

$$F - b = A_2(f_2 - w_0)^2 = A_4(f_4 - w_1)(f_4 - w_2)^3$$

In this case

$$F - b = A_5(f_5 - d_1)^3(f_5 - d_2)^2$$

or

$$F - b = A_5(f_5 - d_1)(f_5 - d_2)^4$$

In the first place we consider the first case. As in the above

$$C(X - 1)(X + 1)^3/ X^2$$
$$= Y^3(Y^2 + A), \qquad A = d_1 - d_2 \neq 0$$

In this case there are two D_k, for which

$$Y^5 + A Y^3 + D_k = (Y - \beta_{k1})(Y - \beta_{k2})(Y - \beta_{k3})(Y - \beta_{k4})^2$$

$$\beta_{14} = i \sqrt{15 A}/5, \qquad \beta_{24} = - \beta_{14}$$

Further

$$(X^4 + 2 X^3 + \alpha_k X^2 - 2 X - 1)/ X^2$$
$$= (X - \alpha_{k1})(X - \alpha_{k2})(X - \alpha_{k3})^2$$
$$\alpha_{13} = \omega, \ \alpha_{23} = \omega^2, \quad \omega = (1 + \sqrt{3}\, i)/2$$

Then $\alpha_{11}, \alpha_{12}, \alpha_{21}, \alpha_{22}$ are different with each other. Let R be the Riemann surface (X, Y) defined by

$$Y^3 (Y^2 + A) = C(X - 1)(X + 1)^3/ X^2$$

On $\alpha_{11}, \alpha_{12}, \alpha_{21}, \alpha_{22}$ there are branch points of order 2 and on ∞ and on 0, 1 there are branch points of order 5 and of order 3, respectively. Hence by the Riemann-Hurwitz relation

the genus of R = 2

Thus by Lemma 3 we have a contradiction.

We next consider the latter case. Then

$$Y(Y^2 + A)^2 = C(X - 1)(X + 1)^3/ X^2$$

There are four D_k for which

$$Y(Y^2 + A)^2 + D_k = (Y - \beta_{k1})(Y - \beta_{k2})(Y - \beta_{k3})(Y - \beta_{k4})^2$$

however there are only two α_k for which

$$(X - 1)(X + 1)^3 + \alpha_k X^2$$

$$= (X - \alpha_{k1})(X - \alpha_{k2})(X - \alpha_{k3})^2 \qquad \alpha_k \neq 0$$

This gives easily a contradiction.

We now consider the case

$$F - b = A_2(f_2 - w_o)^2 = A_4(f_4 - w_1)^4$$

Then

$$F - b = A_5(f_5 - d_1)(f_5 - d_2)^4, \qquad d_1 \neq d_2$$

or

$$= A_5 (F_5 - d_1)^5$$

Assume that the latter is the case. Then F - b has only zeros of order 20 p, where p is a natural number. Let us consider m = 8. Then we have only one possibility $F - b = A_8(f_8 - \alpha)^8$. This implies that F - b has only zeros of order 40 p. Continue this process. Then F - b has only zeros of an arbitrarily high order, which is absurd. Assume that the former case occurs. Then we may put $f_5 - d_1 = L^4$. Hence

$$A_4^{1/4}(f_4 - w_1) = A_5^{1/4} L(L^4 + d_1 - d_2)$$

Let us consider the case m = 8. Two cases may occur:

$$F - b = A_8(f_8 - \alpha_1)^2(f_8 - \alpha_2)^2(f_8 - \alpha_3)^4$$

or

$$= A_8(f_8 - \alpha_1)^4(f_8 - \alpha_2)^4$$

If the former is the case, then

$$L(L^4 + d_1 - d_2) = C(X^4 + d\,X^3 - d\,X - 1)/\,X^2$$

where

$$f_8 - \alpha_1 = g^2, \qquad f_8 - \alpha_2 = h^2$$

$$g = \tilde{E}\,(e^K + e^{-K}), \qquad h = \tilde{E}\,(e^K - e^{-K}), \qquad \tilde{E} = \sqrt{\alpha_2 - \alpha_1}\,/2$$

$$X = e^{2K}$$

In this case the Riemann surface (X, L) is of genus two. This leads to a contradiction similarly. If the latter case occurs, then

$$L(L^4 + d_1 - d_2) = C(f_8 - \alpha_1)(f_8 - \alpha_2)$$

There is one α for which

$$C(f_8 - \alpha_1)(f_8 - \alpha_2) + C\alpha = C(f_8 - \delta)^2, \qquad \alpha \neq 0$$

This gives

$$L(L^4 + d_1 - d_2) + \alpha = (L - \beta_1)(L - \beta_2)^2(L - \beta_3)^2$$

However this implies either $\beta_1 = 0$ or $\beta_2 = 0$ or $\beta_3 = 0$ and $\alpha = 0$, which is absurd.

The third step. There remains only one case

$$F - b = A_2(f_2 - w_0)^2 = A_4(f_4 - w_1)^2(f_4 - w_2)^2$$

$$= A_5(f_5 - d_1)(f_5 - d_2)^2(f_5 - d_3)^2$$

Let Y^2 be $f_5 - d_1$. Then

$$A_2^{1/2}(f_2 - w_0) = A_4^{1/2}(f_4 - w_1)(f_4 - w_2)$$

$$= A_5^{1/2}(Y\,(Y^2 + d_1 - d_2)(Y^2 + d_1 - d_3))$$

Let C be $(w_1 - w_2)^2/4$. Then

$$A_4^{1/2}(f_4 - w_1)(f_4 - w_2) + A_4^{1/2}C$$

$$= A_4^{1/2}(f_4 - u)^2, \qquad u = (w_1 + w_2)/2$$

Further

$$A_5^{\frac{1}{2}}\{Y(Y^2+d_1-d_2)(Y^2+d_1-d_3)+A_4^{\frac{1}{2}}C/A_5^{\frac{1}{2}}\}$$
$$=A_5^{\frac{1}{2}}(Y-\gamma_1)(Y-\gamma_2)^2(Y-\gamma_3)^2$$

In this case

$$\gamma_1+2\gamma_2+2\gamma_3=0$$
$$2\gamma_2\gamma_3^2+2\gamma_2^2\gamma_3+4\gamma_1\gamma_2\gamma_3+\gamma_1\gamma_2^2+\gamma_1\gamma_3^2=0$$
$$\gamma_1\gamma_2^2\gamma_3^2=D$$
$$4\gamma_2\gamma_3+\gamma_2^2+\gamma_3^2+2\gamma_1\gamma_2+2\gamma_1\gamma_3=A+B$$
$$\gamma_2^2\gamma_3^2+2\gamma_1\gamma_2^2\gamma_3+2\gamma_1\gamma_2\gamma_3^2=AB$$

with

$$D=(A_4/A_5)^{\frac{1}{2}}C,\qquad A=d_1-d_2,\qquad B=d_1-d_3$$

The above equation gives either $\gamma_2+\gamma_3=0$ or $\gamma_2^2+3\gamma_2\gamma_3+\gamma_3^2=0$ The former implies $\gamma_1=0$ and $D=0=C$, which is absurd. The latter implies $\gamma_2\gamma_3=-\gamma_1^2/4$. Then $A+B=-5\gamma_1^{2/4}$, $AB=5\gamma_1^4/16$ and $(A+B)^2=5AB$, $D=AB\gamma_1/5$. This analysis gives that there are two D_k such that $D_2=D_1$ and

$$f_2-x_1=M_1^2=(A_4/A_2)^{\frac{1}{2}}(f_4-E)^2,\qquad E=(w_1+w_2)/2$$
$$f_2-x_2=NM_2^2=(A_5/A_2)^{\frac{1}{2}}(f_5-\gamma_1')(f_5-\gamma_2')^2(f_5-\gamma_3')^2$$
$$=(A_4/A_2)^{\frac{1}{2}}\{(f_4-w_1)(f_4-w_2)-C\}$$

Here N is an entire function having only simple zeros. Now we can make use of our result in [2]. Let f be the entire function satisfying

$$f+1=a_1(f_2-x_1),\qquad f-1=a_1(f_2-x_2),\qquad a_1(x_2-x_1)=2$$

Then $f^2-a_1^2NM_1^2M_2^2=1$ and we put

$$\Theta(z)=\frac{1}{i}a_1\sqrt{N}M_1M_2f+2ia_1\int_{\alpha_1}^{z}\sqrt{N}M_1M_2f'\,dz$$

where α_1 is a zero of N and the integral is taken along a path connecting

α_1 with z. Of course it depends on paths. Then

$$f(z) = \cos\theta(z), \qquad f_2 - w_0 = \frac{x_2 - x_1}{2}\cos\theta$$

The fourth step. Assume that for some $n \geqq 2$, $p = 2^n$

$$F - b = A_2(f_2 - w_0)^2 = A_4(f_4 - w_1)^2(f_4 - w_2)^2$$

$$= A_p\,(f_p - \beta_1)(f_p - \beta_2)\prod_3^{s_p}(f_p - \beta_j)^{2\nu_j}, \qquad 2 + 2(\nu_3 + \ldots + \nu_{s_p}) = p$$

Then by Lemma in [2], the fifth step F - b = A cos K - A, which is a part of our final result. Hence our original problem P(F) reduces to the following simultaneous equations

$$F - b = A_a(f_2 - w_0)^2 = A_4(f_4 - w_1)^2(f_4 - w_2)^2$$

$$= A_5(f_5 - d_1)\,(f_5 - d_2)^2(f_5 - d_3)^2$$

$$= A_p\prod_1^{s_p}(f_p - \beta_j)^{2\,\nu_j}$$

$$2\sum_1^{s_p}\nu_j = p \equiv 2^n$$

for every integer $n \geqq 3$. Hence with $Y^2 = f_5 - d_1$

$$A_2^{\frac{1}{2}}(f_2 - x_1) = A_4^{\frac{1}{2}}(f_4 - E)^2 \qquad E = (w_1 + w_2)/2$$

$$= A_5^{\frac{1}{2}}(Y - \gamma_1)(Y - \gamma_2)^2(Y - \gamma_3)^2$$

$$= A_p^{\frac{1}{2}}(f_p - \beta_1') \cdots (f_p - \beta_{p/2}')$$

Firstly we consider the case p = 8. Then the last expression has one of the following forms:

$$A_8^{\frac{1}{2}}(f_8 - \beta_1')(f_8 - \beta_2')(f_8 - \beta_3')^2$$

$$A_8^{\frac{1}{2}}(f_8 - \beta_1')^2(f_8 - \beta_2')^2$$

$$A_8^{\frac{1}{2}}(f_8 - \beta_1')(f_8 - \beta_2')^3$$

The first and the third cases had been treated in the second step and there had appeared an exceptional case, for which $f_2 - x_1 = A\cos K - A$.

This occurs from the first case. We now reexamine the situation. Let us put

$$(f_4 - w_1)(f_4 - w_2) + D = (f_4 - E)^2, \qquad E = (w_1 + w_2)/2$$

Then

$$A_5^{\frac{1}{2}} \{Y(Y^2 + d_1 - d_2)(Y^2 + d_1 - d_3) + (A_4/A_5^{\frac{1}{2}}D\}$$

$$= A_5^{\frac{1}{2}} (Y - \gamma_1)(Y - \gamma_2)^2(Y - \gamma_3)^2$$

This is equal to

$$A_4^{\frac{1}{2}} (f_4 - E)^2$$

In this case we have

$$2\, d_1 - d_2 - d_3 = -\, 5\, \gamma_1^{\;2}/4$$

$$(d_1 - d_2)(d_1 - d_3) = 5\, \gamma_1^{\;4}/16$$

$$(A_4/A_5)^{\frac{1}{2}}\, D = \gamma_1^{\;5}/16$$

Hence there are two D satisfying the above relations, that is,

$$(A_4/A_5)^{\frac{1}{2}}D = \pm \frac{2i}{5} (d_1 - d_2)(d_1 - d_3) \sqrt{2d_1 - d_2 - d_3}$$

Therefore

$$A_2^{\frac{1}{2}} (f_2 - x_2) = A_5^{\frac{1}{2}} (Y - \gamma_1{}')(Y - \gamma_2{}')^2(Y - \gamma_3{}')^2$$

This implies that

$$A_2^{\frac{1}{2}} (f_2 - x_2) = A_8^{\frac{1}{2}} (f_8 + \beta_1{}'')^2(f_8 + \beta_2{}'')^2$$

since $f_8 + \beta_1{}'$ and $f_8 + \beta_2{}'$ have only zeros of even order. Thus we have

$$(f_8 + \beta_1)(f_8 + \beta_2)(f_8 + \beta_3)(f_8 + \beta_4)$$

$$= (f_8 + \beta_1{}')(f_8 + \beta_2{}')(f_8 + \beta_3{}')^2 - C$$

$$= (f_8 + \beta_1{}'')^2(f_8 + \beta_2{}'')^2 + C$$

where $C = (A_4/A_8)^{\frac{1}{2}} D$. In this case

$$2\beta_3' = \beta_1' + \beta_2' = \beta_1'' + \beta_2''$$

$$-C = (\beta_1' - \beta_2')^4/128 = (\beta_1'' - \beta_2'')^4/32$$

Thus

$$-A_4^{\frac{1}{2}} (w_1 - w_2)^2/4 = A_8^{\frac{1}{2}}(\beta_1' - \beta_2')^4/128$$

We now put

$$f_8 + \beta_1' = g^2, \qquad f_8 + \beta_2' = h^2$$

$$g + h = \sqrt{\beta_1' - \beta_2'}\, e^L, \qquad g - h = \sqrt{\beta_1' - \beta_2'}\, e^{-L}$$

Then

$$(f_8 + \beta_1')(f_8 + \beta_2')(f_8 + \beta_3')^2$$

$$= (\beta_1' - \beta_2')^4(e^{8L} - 2 + e^{-8L})/4^4$$

Hence

$$A_2^{\frac{1}{2}} (f_2 - w_o) = A_4^{\frac{1}{2}} (f_4 - w_1)(f_4 - w_2)$$

$$= A_4^{\frac{1}{2}} (f_4 - E)^2 - A_4^{\frac{1}{2}} (w_1 - w_2)^2/4$$

$$= A_8^{\frac{1}{2}} (f_8 + \beta_1')(f_8 + \beta_2')(f_8 + \beta_3')^2 + 2\, A_8^{\frac{1}{2}} (\beta_1' - \beta_2')^4/4^4$$

$$= A_8^{\frac{1}{2}} (\beta_1' - \beta_2')^4(e^{8L} + e^{-8L})/4^4$$

This implies that

$$F - b = A_2(f_2 - w_o)^2$$

$$= A_8\, (\beta_1' - \beta_2')^8(e^{16L} + e^{-16L} + 2)/4^8$$

$$= A(\cos K + 1), \qquad Ki = 16L$$

This is just a part of our final result.

Next we consider the case $p = 2^n$. Then

$$A_2^{\frac{1}{2}} (f_2 - x_1) = A_4^{\frac{1}{2}} (f_4 - E)^2$$

$$= A_8^{\frac{1}{2}} (f_8 - \beta_1'^{(3)})^2(f_8 - \beta_2'^{(3)})^2$$

$$= A_5^{\frac{1}{2}} (Y - \gamma_1)(Y - \gamma_2)^2 (Y - \gamma_3)^2$$

$$= A_p^{\frac{1}{2}} (f_p - \beta_1{}'^{(n)})(f_p - \beta_2{}'^{(n)}) \prod_{j=3}^{s} (f_p - \beta_j{}'^{(n)})^2$$

In this case

$$A_2^{\frac{1}{2}} (w_0 - x_1) = A_4^{\frac{1}{2}} (w_2 - w_1)^2/4$$

As in the case n = 3, we have

$$A_2^{\frac{1}{2}} (f_2 - x_2) = A_4^{\frac{1}{2}} \{(f_4 - w_1)(f_4 - w_2) - (w_1 - w_2)^2/4\}$$

$$= A_5^{\frac{1}{2}} (Y - \gamma_1{}')(Y - \gamma_2{}')^2 (Y - \gamma_3{}')^2$$

$$= A_8^{\frac{1}{2}} (f_8 - \beta_1{}''^{(3)})(f_8 - \beta_2{}''^{(3)})(f_8 - \beta_3{}''^{(3)})^2$$

Hence

$$A_4^{\frac{1}{2}} \frac{(w_2 - w_1)^2}{4} = \frac{1}{32} A_8^{\frac{1}{2}} (\beta_1{}'^{(3)} - \beta_2{}'^{(3)})^4$$

Further by Lemma in [2], the fifth step

$$A_4^{\frac{1}{4}} (f_4 - E) = A_p^{\frac{1}{4}} G (X^{p/4} - X^{-p/4})$$

$$= A_p^{\frac{1}{4}} G (X^{p/4} + 2i - X^{-p/4}) - 2i A_p^{\frac{1}{4}} G$$

where

$$X = e^{2L}, \qquad f_p - \beta_1{}'^{(n)} = T^2, \qquad f_p - \beta_2{}'^{(n)} = S^2$$

$$T + S = \sqrt{\beta_2{}'^{(n)} - \beta_1{}'^{(n)}}\, e^L, \qquad G = (\beta_2{}'^{(n)} - \beta_1{}'^{(n)})^{p/4}/4^{p/4}$$

$$E = (w_1 + w_2)/2$$

The left hand side expression of the above equations is equal to

$$A_8^{\frac{1}{4}} (f_8 - (\beta_1{}'^{(3)} + \beta_2{}'^{(3)})/2)^2 - A_8^{\frac{1}{4}} (\beta_2{}'^{(3)} - \beta_1{}'^{(3)})^2/4$$

Hence

$$A_8^{\frac{1}{2}} (\beta_2{}'^{(3)} - \beta_1{}'^{(3)})^4/16 = -4 A_p^{\frac{1}{2}} G^2$$

Therefore

$$A_2^{\frac{1}{2}}(f_2 - w_o) = A_p^{\frac{1}{2}} G^2 (X^{p/2} - 2 + X^{-p/2}) + 2 A_p^{\frac{1}{2}} G^2$$
$$= A_p^{\frac{1}{2}} G^2 (X^{p/2} + X^{-p/2})$$

By $X = e^{2L}$

$$F - b = 2 A_p G^4 (\cos K + 1), \qquad Ki = 2pL$$

This is a part of our final result.

Therefore we may assume that $P(f_2)$ is just

$$A_2^{\frac{1}{2}}(f_2 - x_1) = A_4^{\frac{1}{2}}(f_4 - E)^2$$
$$= A_5^{\frac{1}{2}}(Y_1 - \gamma_1)(Y_1 - \gamma_2)^2(Y_1 - \gamma_3)^2$$
$$= A_p^{\frac{1}{2}} \prod_1^{p/4} (f_p - \beta_j'^{(n)})^2$$

This gives

$$A_4^{\frac{1}{4}}(f_4 - x_1') = A_8^{\frac{1}{4}}(f_8 - E_3)^2 \qquad 2E_3 = (\beta_1'^{(3)} + \beta_2'^{(3)})$$
$$= A_5^{\frac{1}{4}}(Y_2 - \gamma_1^{(2)})(Y_2 - \gamma_2^{(2)})^2(Y_2 - \gamma_3^{(2)})^2$$
$$= A_p^{\frac{1}{4}} \prod_1^{p/4} (f_p - \alpha_j)$$

Assume that the last expression for $p = 2^n$ has the form

$$A_p^{\frac{1}{4}}(f_p - \beta_1''^{(n)})(f_p - \beta_2''^{(n)}) \prod_3^{s} (f_p - \beta_j''^{(n)})^2$$

This gives

$$A_8^{\frac{1}{4}}(\beta_1'^{(3)} - \beta_2''^{(3)})^2/4 = -2 A_p^{\frac{1}{4}} (\beta_2''^{(n)} - \beta_1''^{(n)})^{p/4}/4^{p/4}$$

Further the following equality is already known

$$A_4^{\frac{1}{2}}(w_2 - w_1)^2/4 = A_8^{\frac{1}{2}}(\beta_2'^{(3)} - \beta_1'^{(3)})^4/32$$

Let us put

$$T^2 = f_p - \beta_1''^{(n)}, \qquad S^2 = f_p - \beta_2''^{(n)}$$
$$T + S = \sqrt{\beta_2''^{(n)} - \beta_1''^{(n)}}\; e^L, \qquad X = e^{2L}$$

Then

$$A_4^{1/4}(f_4 - E) = A_p^{1/4}(\beta_2''^{(n)} - \beta_1''^{(n)})^{p/4}(X^{p/4} + X^{-p/4})/4^{p/4}$$

Further

$$A_2^{1/2}(f_2 - w_o) = A_p^{1/2}(\beta_2''^{(n)} - \beta_1''^{(n)})^{p/2}(X^{p/2} + X^{-p/2})/4^{p/2}$$

and hence

$$\begin{aligned} F - b &= A_2(f_2 - w_o)^2 \\ &= A_p\,(\beta_2''^{(n)} - \beta_1''^{(n)})^p\,(X^p + X^{-p} + 2)/\,4^p \\ &= A \cos K + A, \qquad Ki = 2pL \end{aligned}$$

This is a part of our final result. Therefore we may assume that there does not appear any exceptional case. So we can proceed a step further and get $P(f_4)$. This process can be continued ad infinitum. Each step contains some exceptional cases. However each exceptional case gives a part of our final result. Its proof is quite similar as in the above. So we shall omit its proof.

The fifth step. With some simplified notations and $p = 2^n$

$$\begin{aligned} A_p^{1/p}(f_p - E_n) &= A_{2p}^{1/p}(f_{2p} - \beta_1^{(n+1)})(f_{2p} - \beta_2^{(n+1)}) \\ &= A_5^{1/p}\,Y_n(Y_n^2 - \gamma_1^{(n)})(Y_n^2 - \gamma_2^{(n)}) \\ &= A_{4p}^{1/p}\prod_{j=1}^{4}(f_{4p} - \beta_j'^{(n+2)}) \end{aligned}$$

where

$$E_n = (\beta_1^{(n)} + \beta_2^{(n)})/2$$

This is just a part of $P(f_p)$. Therefore

$$\begin{aligned} A_p^{1/p}(f_p - x_1^{(n)}) &= A_{2p}^{1/p}(f_{2p} - E_{n+1})^2 \\ &= A_5^{1/p}\,(Y_n + \gamma_1'^{(n)})(Y_n + \gamma_2'^{(n)})^2(Y_n + \gamma_3'^{(n)})^2 \\ &= A_{4p}^{1/p}(f_{4p} - \beta_1^{(n+2)})^2(f_{4p} - \beta_2^{(n+2)})^2 \end{aligned}$$

Further

$$A_p^{1/p}(f_p - x_2^{(n)})$$

$$= A_{2p}^{1/p}\{(f_{2p} - \beta_1^{(n+1)})(f_{2p} - \beta_2^{(n+1)}) - \frac{(\beta_1^{(n+1)} - \beta_2^{(n+1)})^2}{4}\}$$

$$= A_5^{1/p}(Y_n + \gamma_1''^{(n)})(Y_n + \gamma_2''^{(n)})^2(Y_n + \gamma_3''^{(n)})^2$$

$$= A_{4p}^{1/p}(f_{4p} - \beta_1''^{(n+2)})(f_{4p} - \beta_2''^{(n+2)})(f_{4p} - \beta_3''^{(n+2)})^2$$

Hence

$$A_{2p}^{1/p}\ \frac{(\beta_1^{(n+1)} - \beta_2^{(n+1)})^2}{4} = \frac{1}{32} A_{4p}^{1/p}\ (\beta_1^{(n+2)} - \beta_2^{(n+2)})^4$$

This is a key relation in our final step. Let us consider

$$f_2 - x_1 = (A_4/A_2)^{\frac{1}{2}}\ (f_4 - E)^2$$

which is the same as

$$f_2 - w_o = (A_4/A_2)^{\frac{1}{2}}\ \frac{(w_1 - w_2)^2}{4}\ \ (2f_4{}^{*2} - 1)$$

$$f_4{}^* = \frac{\sqrt{2}}{w_1 - w_2}\ \ (f_4 - E)$$

By

$$2(x_2 - x_1) = (A_4/A_2)^{\frac{1}{2}}\ (w_1 - w_2)^2$$

and by

$$f_2 - w_o = \frac{x_2 - x_1}{2}\ \cos\theta$$

we have

$$f_4{}^* = \cos(\theta + 2\pi j)/2, \qquad j = 0,\ 1$$

By induction we can prove that

$$f_{2p}{}^* = \cos\frac{\theta + 2\pi j}{p}, \qquad j = 0,\ 1,\ \ldots,\ p - 1,\ p = 2^n$$

where

$$f_{2p}{}^* = \frac{\sqrt{2}}{\beta_1^{(n+1)} - \beta_2^{(n+1)}}\ (f_{2p} - E_{n+1})$$

Here we need the key relation in order to prove the above relation.

Let $\{\alpha_j\}$ be the set of zeros of N, which appeared in the third step.

Let C be a cycle in $\{|z|<\infty\} - \{\alpha_j\}$, which bounds only two points α_1, α_2 of $\{\alpha_j\}$. Assume that the period of Θ along C is equal to $2q\pi i \neq 0$. Then the period of $(\Theta + 2\pi j)/p$ along the cycle is equal to $2q\pi i/2^n$. Thus

$$f_{2p}^* = \cos\frac{\theta + 2\pi j}{p}$$

is not one-valued along C, which is a contradiction, since f_{2p}^* is one-valued in any finite plane. Therefore

$$\int_{\alpha_1}^{\alpha_2} \sqrt{N}\ M_1 M_2 f'\, dz = 0$$

This is true for every $\{\alpha_j\}$ and for every path of integration and hence for every cycle. Thus

$$\int_{\alpha_1}^{z} \sqrt{N}\, M_1 M_2 f'\, dz = \sqrt{N}\, S$$

with an entire function S. Then $\Theta = \sqrt{N}\, K$ with an entire function K. Therefore

$$F - b = A\,(\cos 2\sqrt{N}\, K + 1)$$

with a suitable constant A, which is just our desired result.

By the above method we can improve Theorem B.

THEOREM 2. Suppose that (*) holds for $m = 3^j (j = 1, 2,\ldots)$ and for $m = 4$. Then the same conclusion as in Theorem A remains true.

We shall not give any proof here. Lemma 3 and the following fact play the role in the proof:

Any closed Riemann surface of genus one is not uniformizable by any pair of entire functions.

REFERENCES

1. R. Nevanlinna, *Le théorème de Picard-Borel et la théorie de fonctions méromorphes*, Gauthier-Villars, Paris, 1929.

2. M. Ozawa, On a characterization of the exponential function and the cosine function by factorization; *Kodai Math. J.*, *1* (1978), 45-74.

3. M. Ozawa, On a characterization of the exponential function and the cosine function by a factorization, III, *Kodai Math. J.*, *2* (1979), 200-210.

4. H. L. Selberg, Algebroid funktionen und umkehrfunktionen abelscher integrale, *Avh. Norske Vid. Akad. Oslo*, *8* (1934), 1-72.

ON THE PRIMENESS OF THE PAINLEVÉ TRANSCENDENTS

Norbert Steinmetz

Mathematisches Institut I
Universität Karlsruhe
Karlsruhe, West Germany

1. INTRODUCTION

Let h be a transcendental meromorphic function in the complex plane. We say that h has f and g as left and right factors, respectively, if

$$h(z) = f(g(z)) \tag{*}$$

Here it is assumed that either f is transcendental meromorphic and g is entire or that f is rational and g is transcendental meromorphic. h is said to be prime, if no nontrivial factorization of type (*) is possible, i.e., if (*) implies that either f or g is a linear function (for notations see, e.g., Ref. [2]).

There is a long list of papers dealing with prime and pseudo-prime functions h (i.e., functions whose factors are not simultaneously transcendental).

In our paper we are concerned with the so-called Painlevé transcendents which are defined implicitly to be solutions of the first two Painlevé differential equations

$$w'' = z + 6w^2 \tag{I}$$

and

$$w'' = a + zw + 2w^3 \quad . \tag{II}$$

It is well known ([4], p. 439) that every solution of (I) or (II) is a transcendental meromorphic function in the plane. Our main result is

THEOREM: The Painlevé transcendents are prime.

1. Notations and preliminary results

We will assume that the reader is familiar with the basic notations and results of Nevanlinna theory (see, e.g., Refs. [3] and [5]). One famous and important result of this theory will be frequently used:

LEMMA 1. Let f be a meromorphic function of finite order. Then,

$$m(r,\frac{f'}{f}) = 0\ (\log r) \ ,$$

as $r \to +\infty$.

We need some properties of the Painlevé transcendents which are denoted by the letter w. If we want to distinguish between the solutions of (I) or (II), we write w_1 or w_2, respectively. Most of the following results can be found in Ref. [8].

LEMMA 2. The Painlevé transcendents are meromorphic functions of finite order with infinitely many poles. More accurately, $m(r,w) = 0(\log r)$, as $r \to +\infty$.

Let z_ν be a pole of w_1 or w_2. Then it is easily found that

$$w_1(z) = \frac{1}{(z-z_\nu)^2} - \frac{z_\nu}{10}(z-z_\nu)^2 - \frac{1}{6}(z-z_\nu)^3 + c_4\ (z-z_\nu)^4 + \ldots \tag{1}$$

or

$$w_2(z) = \frac{\varepsilon_\nu}{z-z_\nu} - \frac{\varepsilon_\nu z_\nu}{6}\ (z-z_\nu) - \frac{1}{4}(\varepsilon_\nu + a)(z-z_\nu)^2 + c_3(z-z_\nu)^3 + \ldots \tag{2}$$

($\varepsilon_\nu = \pm 1$) in a neighbourhood of z_ν. The coefficients c_4 and c_3 remain undetermined.

To some extent the Painlevé transcendents are characterized by their Laurent expansions (1) and (2).

LEMMA 3. Let w be a transcendental meromorphic function of finite order. Assume that $m(r,w) = O(\log r)$ and that there is a polynomial $a(z)$ such that at every pole the Laurent expansion

$$w(z) = \frac{1}{(z-z_\nu)^2} - \frac{a(z_\nu)}{10}(z-z_\nu)^2 + \dots \qquad (3)$$

is valid. Then w satisfies the Painlevé differential equation

$$w'' = a(z) + 6w^2 \qquad (4)$$

and $a(z)$ is necessarily linear (or constant).

Proof. It is easily seen, using (3), that $d := w'' - 6w^2 - a$ vanishes at $z = z_\nu$ and is therefore an entire function. By Lemma 1 and the hypothesis on w it follows that $m(r,d) = O(\log r)$. Hence, d is a polynomial which vanishes at every pole of w and so identically. This proves the first part of Lemma 3. To prove the second part, we shall show that $a''(z)$ vanishes at every pole of w. If $w(z) = \sum_{n=-2}^{\infty} c_n(z-z_\nu)^n$ near z_ν, an easy computation gives $c_{-2} = 1$, $c_{-1} = c_0 = c_1 = 0$, $c_2 = \frac{a(z_\nu)}{10}$, $c_3 = -\frac{a'(z_\nu)}{6}$, while c_4 satisfies $12c_4 = 12c_4 + \frac{1}{2}a''(z_\nu)$. This proves Lemma 3.

There is a similar result for meromorphic functions with the Laurent expansion

$$w(z) = \frac{\varepsilon_\nu}{z-z_\nu} - \frac{1}{6}\varepsilon_\nu b(z_\nu)(z-z_\nu) - \frac{1}{4}(\varepsilon_\nu b'(z_\nu)+a(z_\nu))(z-z_\nu)^2 + \dots \qquad (5)$$

near every pole. Here, $\varepsilon_\nu = \underline{+}1$ and $a(z)$ and $b(z)$ are polynomials.

LEMMA 4. Let w be a transcendental meromorphic function of finite order. Assume that $m(r,w) = O(\log r)$ and that w has the Laurent expansion (5) at every pole $z = z_\nu$. Then there are exactly two cases as follows:

(α) w satisfies the Painlevé differential equation

$$w'' = a(z) + b(z)w + 2w^3 , \qquad (6)$$

where a is constant and b is linear (or constant);

(β) w satisfies the Riccati differential equation

$$w' = -\varepsilon\{\frac{b(z)}{2} + w^2\} \tag{7}$$

Here $\varepsilon = +1$ is constant, and we have $\varepsilon_\nu = \varepsilon$ for all ν. The polynomial $a(z)$ is given by $a(z) = -\frac{1}{2}\varepsilon b'(z)$.

Proof. As in the proof of Lemma 3 it is shown that $w''-a(z)-b(z)w-2w^3$ is a polynomial which vanishes at every pole of w and therefore identically. If one tries to compute the Laurent coefficients of $w(z) = \sum_{n=-1}^{\infty} c_n(z-z_\nu)^n$, one easily gets the following equations (cf. [8]): $c_{-1}^2 = 1$, $c_0 = 0$, $b(z_\nu)c_{-1} + 6c_1 = 0$, $2c_2 = a(z_\nu) + b'(z_\nu)c_{-1} + 6c_2$, $6c_3 = a'(z_\nu)+b(z_\nu)c_1 + \frac{1}{2}b''(z_\nu)c_{-1} + 6c_{-1}c_1^2 + 6c_3$. The last equation gives $a'(z_\nu)+\frac{1}{2}\varepsilon_\nu b''(z_\nu)=0$. If $c_{-1} =$ takes both values +1 and -1 infinitely often, then we conclude $a'(z) \equiv b''(z) \equiv 0$, as stated in part ($\alpha$).

Now suppose that ε_ν has the constant value ε for $\nu \geq \nu_0$. Then it is easily shown that $w' + \varepsilon\frac{b(z)}{2} + w^2$ is a rational function which vanishes at $z = z_\nu$ for $\nu \geq \nu_0$ (poles can occur at $z = z_\nu$ for $\nu < \nu_0$). Hence, this rational function must vanish identically and w must be a solution of (7). At every pole w has residue ε, and so $\varepsilon_\nu = \varepsilon$ for all ν. Differentiation of (7) gives

$$w'' = -\varepsilon\frac{b'(z)}{2} + b(z)w + 2w^3$$

which shows, indeed, that $a(z) \equiv -\varepsilon\frac{b'(z)}{2}$.

We need some results on composite functions $h(z) = f(g(z))$. The first one is due to Clunie [1]:

LEMMA 5. Let $h(z) = f(g(z))$ be a transcendental meromorphic function. If f is nonrational, then

$$T(r,g) = O(T(r,h)) \text{ as } r \to +\infty. \tag{8}$$

It is a simple exercise to show that, if $f(u_0) = c$, then $m\{r,\frac{1}{g-u_0}\} \leq m\{r,\frac{1}{h-c}\} + O(1)$ (see [3], p. 54). This is also true if $c=\infty$ or $u_0 = \infty$. What we need is

LEMMA 6. Let $h(z) = f(g(z))$ be a meromorphic function and let u_0 be a pole of f. Then

$$m\{r, \frac{1}{g-u_0}\} \leq m(r,h) + O(1)$$

if u_o is finite, and

$$m(r,g) \le m(r,h) + O(1)$$

if $u_o = \infty$

2. PROOF OF THE THEOREM

The proof of the Theorem is cut into two pieces, Case A and Case B. We start with

Case A, where the left factor is transcendental.

Assume that w is a Painlevé transcendent which can be written as

$$w(z) = f(g(z)) \tag{9}$$

where f is a transcendental meromorphic function. Hence g is entire transcendental or a polynomial. f must have infinitely many poles $u_1, u_2, \ldots$. If there were only a finite number of poles, $u_1, u_2, \ldots, u_p$, say, then

$$\bar{N}(r,w) \le \sum_{j=1}^{p} N\{r, \frac{1}{g-u_j}\} \le pT(r,g) + O(1) = o(T(r,w))$$

by Lemma 5, this contradicts Lemma 2. From Lemma 6 it follows that g takes every value u_j. If $w = w_1$ is a solution of the first Painlevé differential equation, it has only double poles. So the poles of f with the exception of at most two are double poles. We may assume that u_1, u_2 and u_3 are double poles of f,

$$f(u) = \frac{A_j}{(u-u_j)^2} + \ldots$$

in a neighbourhood of u_j. If $g(z_\nu) = u_j$, then it follows that

$$f(g(z)) = \frac{A_j}{g'^2(z_\nu)} \frac{1}{(z-z_\nu)^2} + \ldots,$$

and so

$$g(z_\nu) = u_j \quad \text{implies} \quad g'^2(z_\nu) = A_j \tag{10}$$

as follows from (1).

If $w = w_2$, is the second Painlevé transcendent, all poles of f are simple,

$$f(u) = \frac{B_j}{u-u_j} + \ldots$$

in a neighbourhood of u_j. If $g(z_\nu) = u_j$, we have

$$f(g(z)) = \frac{B_j}{g'(z_\nu)} \frac{1}{z-z_\nu} + \ldots$$

near z_ν. This gives $g'(z_\nu) = \pm B_j$, hence $g'^2(z_\nu) = B_j^2 =: A_j$. So in both cases (10) holds true. From (10) it follows that

$$H_j(z) := \frac{g'^2(z)-A_j}{g(z)-u_j} \tag{11}$$

is an entire function. The same is true for

$$H_{jk}(z) := \frac{H_j(z)-H_k(z)}{U_J-U_K} = \frac{g'^2(z)-P_{jk}(g(z))}{(g(z)-u_j)(g(z)-u_k)} \tag{12}$$

$(j \neq k)$. Here P_{jk} is a certain linear polynomial which depends on u_j and u_k.

Now assume that g is a (nonlinear) polynomial. Then H_{12} is also a polynomial which vanishes at $z = \infty$ and therefore identically. This gives $g'^2 \equiv P_{12}(g)$ or

$$g(z) = c_o + c_2 (z-z_o)^2 \tag{13}$$

with certain coefficients c_o, $c_2 \neq 0$. Our assertion (9) yields $w(z) = f(c_o + c_2(z-z_o)^2) = \tilde{f}((z-z_o)^2)$, i.e., $w(z + z_o)$ is an even meromorphic function. But no solution of (I) or (II) is even with respect to any z_o. This shows that (9) is impossible if g is a polynomial.

We may now assume that g is transcendental entire. We build up the function

$$H_{123}(z) := \frac{H_{12}(z)-H_{13}(z)}{u_2-u_3} = \frac{g'^2(z)-P_{123}(g(z))}{(g(z)-u_1)(g(z)-u_2)(g(z)-u_3)} \tag{14}$$

Here P_{123} stands for a polynomial of degree ≤ 2. Clearly H_{123} is entire, Since g is of finite order (less or equal to the order of w), it follows from Lemma 1 that

$$m(r, \frac{g'^2}{(g-u_1)(g-u_2)(g-u_3)}) = O(\log r)$$

Since $\left|\frac{P_{123}(u)}{(u-u_1)(u-u_2)(u-u_3)}\right|$ remains bounded if u is bounded away from the values u_1, u_2, u_3, it follows that

$$m(r, \frac{P_{123}(g)}{(g-u_1)(g-u_2)(g-u_3)}) \leq \sum_{j=1}^{3} m(r, \frac{1}{g-u_j}) + O(1)$$

$$\leq 3m(r,w) + O(1) = O(\log r)$$

by Lemma 5 and Lemma 2. Thus we have shown that $T(r,H_{123}) = m(r,H_{123}) = O(\log r)$, i.e., H_{123} is a polynomial. Rewriting equation (14), we are led to the differential equation

$$u'^2 = P_{123}(u) + H_{123}(z)(u-u_1)(u-u_2)(u-u_3) \tag{15}$$

Since (15) has an entire transcendental solution $u = g(z)$, this requires $H_{123}(z) \equiv 0$ by a well known theorem of Wittich [7] on entire solutions of algebraic differential equations.

What we have thus shown is that g is a solution of the differential equation

$$u'^2 = P_{123}(u) \tag{16}$$

where P_{123} is a quadratic polynomial. It is well known and easily proved that every nonconstant solution of (16) is periodic. Hence, it would follow from (9) that w is periodic, while (I) and (II) have no periodic solutions (apart from the trivial solution $w = 0$ of equation (ii), where $a = 0$).

We have shown that the Painlevé transcendents are at least pseudo-prime. What is left is

Case B, where the left factor is a rational function.

We want to show that (9) is impossible, if f is a nonlinear rational function. To do this we may assume that f has a pole at infinity. If this is not the case, we consider the factorization $w(z) = f(u_o + \frac{1}{\tilde{g}(z)} = \tilde{f}(\tilde{g}(z))$, where u_o is a finite pole of f and where $\tilde{f}$ and $\tilde{g}$ are defined by $\tilde{g} = \frac{1}{g-u_o}$ and $\tilde{f}(u) = f(u_o + \frac{1}{u})$. It is advantageous to consider first

Subcase a: The second Painlevé transcendent.

f has a simple pole at $u = \infty$. There is no loss of generality to assume that

$$f(u) = u + \frac{R}{u} + \frac{S}{u^2} + \ldots \tag{17}$$

in a neighbourhood of infinity. For, if $f(u) = Au + B + \frac{C}{u} + \ldots$, $A \neq 0$, (9) yields another factorization $w(z) = F(h(z))$, where $h = Ag + B$ and $F(Au + B) = f(u)$. At infinity F behaves like $u + \frac{CA}{u} + \ldots$.

Now let z_ν be a pole of g (by Lemma 6 g has infinitely many poles). A short computation yields that g has the Laurent development

$$g(z) = \frac{\varepsilon_\nu}{z-z_\nu} - \frac{\varepsilon_\nu}{6}(z_\nu+6R)(z-z_\nu) - \frac{1}{4}(\varepsilon_\nu+a+4S)(z-z_\nu)^2 + \ldots$$

By Lemma 4 g is a solution of the Painlevé differential equation

$$g'' = a + 4S + (z + 6R)g + 2g^3 \tag{18}$$

If we replace the derivative g" in the equation

$$f''(g)g'^2 + f'(g)g'' = a + zf(g) + 2f^3(g) \tag{19}$$

by the right hand side of (18), we get a new differential equation

$$g'^2 = H(g) + zK(g)$$

where H and K are given by

$$H(u) = \frac{a + 2f^3(u) - (a+4S+6Ru+2u^3)f'(u)}{f''(u)} \tag{20}$$

and

$$K(u) = \frac{f(u) - uf'(u)}{f''(u)}$$

But it was shown in [6] that a binomial differential equation of type (2) can only have transcendental meromorphic solutions, if either H or K vanishes identically. Now $K(u) \equiv 0$ implies $\frac{f'(u)}{f(u)} = \frac{1}{u}$ or $f(u) = u$ which is impossible. If H vanishes identically, we have

$$\frac{f'(u)}{2f^3(u)+a} = \frac{1}{a+4S+6Ru+2u^3} \tag{21}$$

We will show without integration that (21) has no rational nonlinear solution. If f has a pole at u_0, the left hand side of (21) vanishes at $u = u_0$, but the right hand side does not if $u_0 \neq \infty$. So f has only one (simple) pole at infinity and is therefore linear.

Subcase b: The first Painlevé transcendent.

In this case f might have a simple pole or a double pole at infinity. Let us first assume that the pole at infinity is simple. As in Subcase a, we may assume that f is given by (17) near infinity. If z_ν is a (necessarily double) pole of g, we will find by comparing coefficients

$$g(z) = \frac{1}{(z-z_\nu)^2} - \frac{1}{10}(z_\nu + 10R)(z-z_\nu)^2 + \ldots$$

By Lemma 3 g satisfies

$$g'' = z + 10R + 6g^2 \tag{22}$$

If we replace in (23)

$$f''(g)g'^2 + f'(g)g'' = z + 6f^2(g)$$

the derivative g" by the right hand side of (22), we are led to the differential equation (20), where now H and K are given by

$$H(u) = \frac{6f^2(u) - (10R+6u^2)f'(u)}{f''(u)}$$

and

$$K(u) = \frac{1 - f'(u)}{f''(u)}$$

As in Subcase a, we must have either $H(u) \equiv 0$ or $K(u) \equiv 0$. The latter is of course impossible, since f is assumed to be nonlinear. From $K(u) \equiv 0$ follows

$$\frac{f'(u)}{f^2(u)} = (\frac{5}{3} R + u^2)^{-1}$$

which gives $f(u) = u$ for $R = 0$ and $f(u) = -(\sigma \arctan \sigma u)^{-1}$ if $R = \frac{3}{5}\sigma^{-2} \neq 0$ in contrast to our hypothesis.

The last possibility is that f has a double pole at infinity. It is no loss of generality to assume

$$f(u) = u^2 + R + \frac{S}{u} + \dots \tag{24}$$

as $u \to \infty$. As in the preceding cases we find at an arbitrary pole of g

$$g(z) = \frac{\varepsilon_\nu}{z-z_\nu} - \frac{1}{2}\varepsilon_\nu R(z-z_\nu) - \frac{S}{2}(z-z_\nu)^2 + \dots \tag{25}$$

$\varepsilon_\nu = \underline{+} 1$. Thus Lemma 4 gives with $a(z) \equiv 2S$ and $b(z) \equiv 3R$

$$g'' = 2S + 3Rg + 2g^3$$

or

$$g'^2 = T + 4Sg + 3Rg^2 + g^4 \tag{26}$$

(T constant). But every solution of (26) is periodic and so is $w(z) = f(g(z))$. This contradiction shows that no factorization (9) is possible.

REFERENCES

1. J. Clunie, The composition of entire and meromorphic functions, in McIntyre Memorial Volume, Ohio Univ. Press, 1970.

2. F. Gross, Factorization of meromorphic functions, Math. Research Center, Naval Research Laboratory, Washington, D.C., 1972.

3. W. K. Hayman, Meromorphic functions, Oxford University Press, Oxford, 1964.

4. E. Hille, Ordinary differential equations in the complex domain, Wiley and Sons, New York, 1976.

5. R. Nevanlinna, Eindeutige analytische Funktionen, Berlin: Springer-Verlag, 1936.

6. N. Steinmetz, Bemerkung zu einem Satz von Yosida. Proc. Coll. Complex Analysis, Joensuu, Finland 1978. Lecture Notes in Mathematics, Vol. 747, Springer-Verlag, 1979.

7. H. Wittich, Ganze transzendente Lösungen algebraischer Differentialgleichungen, *Math. Ann.* 122, 221-234 (1950).

8. H. Wittich, Eindeutige Lösungen der Differentialgleichungen $w'' = P(z,w)$. *Math. Ann.* 125, 355-365 (1963).

ON COMMON RIGHT FACTORS OF F AND $F^{(N)}$

Norbert Steinmetz

Universität Karlsruhe (TH)
Mathematisches Institut 1
75 Karlsruhe 1, Englerstre, 2
Federal Republic of Germany

Chung-Chun Yang

Naval Research Laboratory
Washington, D. C.

1. INTRODUCTION

Let $F(z) = f(g(z))$ where both f and g are nonlinear meromorphic functions and are called left and right factors of F respectively. It has been shown in [1] that a transcendental meromorphic function F and its first derivative F' cannot have a common right nonlinear factor other than the form $e^{az+b} + c$, where a, b, and c are constants. A natural question will be: What can be said about possible common right factors of F and its second derivative F'' or in general, $F^{(n)}$ $(n \geq 3)$? Recently in [2], the simpler problem that finding the possible common right meromorphic factors of F,F'', and $F^{(4)}$ has been dealt with. In this note, we shall study the common right entire factors of F and $F^{(n)}$. We shall prove among other things that the possible common right entire factors are the forms: $c_1 e^{c_2 z} + c_3$ and $c_1 \cos(c_2 z + c_3) + c_4$, where all the c_i are constants.

THEOREM 1. Let F(z) be a transcendental meromorphic function. Assume that $F = f(g)$ and $F^{(n)} = h(g)$ $(n \geq 3)$, where g is transcendental entire, f and h are nonlinear meromorphic functions. Then either f satisfies:
$A_n(z)f^{(n)}(z) + \ldots\ldots A_{j+1}(z)f^{(j+1)}(z) + A_j(z)f^{(j)}(z) \equiv 0$, where all

the A_i are rational functions, or $g(z) = c_1e^{c_2z} + c_3$ or $g(z) = c_1\cos(c_2z + c_3) + c_4$; where all the c_i are constants.

In [8], Wittich showed that the order of any transcendental entire solution of a linear differential equation with polynomials as its coefficients must be positive and finite. From this observation and Theorem 2, we can state the following results.

COROLLARY 1. Let F(z) be a transcendental meromorphic function of finite order. Furthermore, if F is periodic with $F = f(g)$ and $F^{(n)} = h(g)$ $(n \geq 2)$ for some nonlinear entire functions f, h, and g. Then either $F = f(q)$, q a quadratic form or g assumes one of the forms stated in Theorem 2.

COROLLARY 2. Let F be a transcendental meromorphic function of order less than one. Furthermore, F is not pseudo-prime, then F and $F^{(n)}$ $(n \geq 2)$ can not have any common right factor which is transcendental.

2. PRELIMINARY LEMMAS

This section contains several lemmas that will be used in the proofs of our results.

LEMMA 1 [6, Theorem 1]. Let $\{h_0,h_1,\ldots\ldots h_m\}$ and $\{F_0,F_1,\ldots\ldots F_m\}$ be two sets of $m + 1$ $(m \geq 1)$ meromorphic functions with $F_i \not\equiv 0$; $i = 0,1,2,\ldots\ldots m$. Suppose that g is a nonconstant entire function and satisfies:

$$\sum_{i=0}^{m} T(r,h_i) = kT(r,g) + S(r,g)$$

where k is a positive constant and S(r,g) denotes some quantity satisfying $S(r,g) = oT(r,g)$ as $r \to \infty$ outside a set of r values of finite length. If

$$F_0(g)h_0 + F_1(g)h_1 + \ldots\ldots\ldots F_m(g)h_m \equiv 0$$

then there exist polynomials $p_0,p_1,\ \ldots\ldots,p_m$ with $p_i \not\equiv 0$ for $i = 0, 1,, \ldots\ldots m$ such that

$$p_0(g)h_0 + p_1(g)h_1 + \ldots\ldots\ p_m(g)h_m \equiv 0$$

LEMMA 2 [7, Theorem 3]. Let $P(z, w,w',\ldots,w^{(n)})$ be a differential polynomial in w(i.e., a polynomial in $w,w'\ldots w^{(n)}$ with coefficients a(z) satisfying $T(r,a(z)) = S(r,w))$, and H(z,w) be a quotient of entire functions of two variables z and w. Suppose that there exists an admissible solution f(z) for

$$P(z,w,w',\ldots w^{(n)}) = H(z,w) \tag{E}$$

Then H(z,w) is a polynomial in w with degree in w at most d(P) = $\max\{j_0 + 2j_1 + \ldots + (n+1)j_n;\ a(z)w^{j_0}(w')^{j_1}(w^{(2)})^{j_2}\ldots(w^{(n)})^{j_n}$ is a term in $P(z,w,\ldots w^{(n)})\}$. Moreover, if f is entire then the degree of H is at most $\max(j_0 + j_1 + \ldots + j_n)$.

REMARK. A meromorphic function f is said to be an admissible solution to equation (E) if the following two conditions are satisfied:
(1) $T(r,a(z)) = S(r,f)$ for all the coefficients a(z) in $P(z,w\ldots w^{(n)})$,
(2) $T(r,H(z,t)) = S(r,f)$ for a set of t values in the plane having a finite limit point.

LEMMA 3 [4, Theorem 6]. Let $P(z,w\ldots w^{(n)})$ be as in Lemma 2 and let f be an admissible solution to the differential equation: $P(z,w\ldots w^{(n)}) = 0$. If $P(z,0,0,0,0\ldots 0) \not\equiv 0$, then $m(r,\frac{1}{f}) = S(r,f)$.

LEMMA 4 [5]. Let f be a meromorphic function of finite order t. Then

$$\overline{\lim_{r\to\infty}} \frac{m(r,f'/f)}{\log r} = \max(t-1,0)$$

3. PROOFS OF THE MAIN RESULTS

3.1. PROOF OF THEOREM 1

It follows from the assumption that F = f(g) and F'' = h(g), we have

$$f''(g)g'^2 + f'(g)g'' = h(g) \tag{1}$$

We now can apply Lemma 1. It follows that there exist polynomials A(z), B(z), and C(z) with $ABC \not\equiv 0$ such that

$$A(g)g^2 + B(g)g + C(g) = 0 \tag{2}$$

Eliminating g'^2 from (1) and (2), we get

$$[A(g)f''(g) - B(g)f''(g)]g'' = A(g)h(g) + f''(g)C(g) \tag{3}$$

If $A(g)f'(g) - B(g)f''(g) \not\equiv 0$, then from the above identity, we have

$$g'' = [A(g)h(g) + f''(g)C(g)] \,/\, [A(g)f'(g) - B(g)f''(g)]$$
$$= H_1(g)$$

where H_1 is a meromorphic function. By Lemma 2, we conclude immediately that

$$g'' = ag + b \tag{4}$$

where a,b are constants. Substituting this into (1), we have

$$(g')^2 = [h(g) - f'(g)(a(g) + b)] \,/\, f''(g) = H_2(g)$$

where H_2 is a meromorphic function. Again by Lemma 2, we conclude that

$$(g')^2 = t_1g^2 + t_2g + t_3$$

or

$$(g')^2 = t_1(g-s_1)(g-s_2) \tag{5}$$

where $t_i(i = 1,2,3)$ and $s_j(j = 1,2)$ are constants; $t_1 \neq 0$.
Two cases to be dealt with separately: case (1): $s_1 = s_2$ and case (2): $s_1 \neq s_2$. If case (1) holds, then

$$(g')^2 = t_1(g-s_1)^2 \tag{6}$$

It follows that $g' = c(g - s_1)$ and hence $g = c_1e^{c_2z} + c_3$; all the c_i are constants. Now if case (2) holds, then (5) can be rewritten as

$$g'^2 = t_1(g - s)^2 + u \tag{7}$$

where s and u are constants; $u \neq 0$. The general entire solution of equation (7) has the form: $g(z) = s + c_1\cos(c_2z + c_3)$, where $c_1^2 = -u/t_2$, $c_2^2 = -t_1$, and c_3 is an arbitrary constant. This completes the proof when $Af - Bf \not\equiv 0$. Now we have to treat the case that $Af' - Bf'' \equiv 0$, i.e.

$$\frac{f''(w)}{f'(w)} = \frac{A(w)}{B(w)} \tag{8}$$

Two cases will be treated separately: case (a): A(w) is a constant and case (b): A(w) is not a constant. We treat case (a) first. In this case, we may assume without loss of generality that $A(w) \equiv 1$. Then (2) becomes

$$w'^2 + B(w)w'' + C(w) = 0 \tag{9}$$

Set $B(w) = bw^{d_1} + B_1(w)$, $C(w) = cw^{d_2} + C_1(w)$ where d_1, d_2 are the degrees of $B_1(w)$ and $C_1(w)$ respectively. Then by a result of Wittich [8], either $d_1 + 1 = d_2 > 2$, or $\max(d_1 + 1, d_2) = 2$. Suppose that $d_1 + 1 = d_2 > 2$, then by rewriting (9) as

$$w^{d_1}(bw'' + cw) = -w'^2 - B_1(w)w'' - C_1(w) \tag{10}$$

and then applying Clune's result [3,p.68], we have

$$T(r,bw'' + cw) = m(r,bw'' + cw) = S(r,w)$$

or

$$T(r,bg'' + cg) = S(r,g) \tag{11}$$

Now the central-index $v(r)$ of g satisfies $b(v/z)^2 + c(1 + k_1(z)) = k_2(z)$, where $k_1(z)$ and $k_2(z)$ tend to zero as $z \to \infty$, outside possibly a set of $r(= |z|)$ values of finite length, therefore the order of g is no greater than 1. By Lemma 4 we have $S(r,g) = o(\log r)$. It follows that $bg'' + cg$ can only be constant and this leads to the situation that we encountered in case (a). We now treat the situation $\max(d_1 + 1, d_2) = 2$. Then (9) becomes

$$w'^2 + (b_0 + b_1w)w'' + c_0 + c_1w + c_2w^2 \equiv 0 \tag{12}$$

Again by considering the central-index, we derive

$$(v/z)^2 + h_1(1 + h_1(z))(v/z) = h_2(z)$$

where $h_1(z)$ and $h_2(z)$ tend to zero as $|z| \to \infty$, outside a set of $r(= |z|)$ values of finite length. On the other hand, since

$$\frac{f''(w)}{f'(w)} = \frac{A(w)}{B(w)} = \frac{1}{b_1w + b_0} \quad \text{and} \quad \frac{1}{b_1} \text{ is an integer} \neq 0, -1, 1,$$

we conclude that the order of g is no greater than one. If $g(z)$ never vanishes, then we are done. Assume that $g'(z_0) = 0$ for some z_0. Then differentiating the equation:

$$w'^2 + B(w)w'' + C(w) = 0 \tag{13}$$

we get, by setting $z = z_0$,

$$B(w_0)g'''(z_0) = 0$$

where $w_0 = g(z_0)$. Two cases may arise: case (α): $B(w_0) = 0$ and case (β): $B(w_0) \neq 0$. Suppose that case (α) holds. We may assume without loss of generality that $w_0 = 0$, i.e. $B(w) = bw$ for otherwise replacing g by $g - w_0$. Setting $z = z_0$ in (9), we get $C(0) = 0$. Hence (12) becomes

$$w'^2 + bww'' + w(c_1 + c_2 w) = 0 \tag{14}$$

Two subcases will be considered: subcase (α_1): $c_1 = 0$ and subcase (α_2): $c_1 \neq 0$. Under subcase (α_1) and by setting $y = w'/w$ into (14), we have

$$b_1 y' + (1 + b_1)y^2 + c_2 = 0 \tag{15}$$

We note now that y has a simple pole at $z = z_0$ with residue p, where p is an integer ≥ 2. By comparing the coefficients of the term $1/(z-z_0)$, the above equation leads to

$$-pb_1 + (1 + b_1)p^2 = 0 \tag{16}$$

On the otherhand, $1/b_1 = q$ is also an integer $\neq 0, -1, 1$. Then (16) yields $(q + 1)p = 1$, a contradiction. Now we consider subcase (α_2). Since the order of g is no greater than one, we are done if g never vanishes. So we assume $g(z_1) = 0$ for some z_1. Then it follows from (14) that $g'(z_1) = 0$, but $g''(z_1) \neq 0$ (since $c_1 \neq 0$). Thus every zero of g is of multiplicity 2. Hence $g = k^2$ for some entire function k and equation (14) becomes

$$b_1 kk'' + (b_1 + 2)k'^2 + \frac{c_1}{2} + \frac{c_2}{2} k^2 = 0 \tag{17}$$

Differentiation of equation (17) gives $(3b_1 + 4)k'(z_1)k''(z_1) = 0$. Since $3b_1 + 4 \neq 0$ (as $1/b_1$ is an integer) and $k'(z_1) \neq 0$, we conclude that $k''(z_1) = 0$. Therefore $k''/k = H$ is entire and Lemma 4 yields $T(r,H) = o(\log r)$, since the order of k is equal to the order of g, which is no greater than one. Thus H must be a constant and then (17) yields $k'' = d_1 g + d_2$; d_1 and d_2 are constants. Hence from $g = k^2$ we have $g' = wkk'$ and hence $g'^2 = 4k^2k'^2 = 4g(d_1 g + d_2)$. This goes back to case (2) which has been settled already. Now suppose that case (β) holds, that is $B(w_0) \neq 0$. If g' never vanishes then we are done. Therefore we assume that $g'(z_0) = 0$ for some z_0, then from (13) we conclude $g'''(z_0) = 0$ (but $g''(z_0) \neq 0$ by the uniqueness-theorem for $w'' = -\frac{C(w)}{B(w)} - \frac{w'^2}{B(w)}$). Therefore g''/g' is entire and again it has to be a constant. Consequently we have $g'' = ag+b$ for some constants a,b. The assertion follows. Finally, we have to settle case (b): $A(w)$ is not a constant. Again, we may assume without loss of generality that $A(0) = 0$ and shall treat two subcases separately: subcase (b_1): $B(0) = 0$ and subcase (b_2): $B(0) \neq 0$. Suppose that subcase (b_1) holds. Clearly we may assume that A,B, and C have no common factor. From $A(0) = 0$ and $B(0) = 0$, we must have $C(0) \neq 0$. Applying Lemma 3 to

$$P(z,w,w',w'') = A(w)w'^2 + B(w)w'' + C(w) = 0$$

where $P(z,0,0,0) = C(0) \neq 0$, we have

$$m(r,1/g) = S(r,g) \tag{18}$$

On the otherhand, from $C(0) \neq 0$, $A(0) = 0$ and $B(0) = 0$ it follows that $g(z)$ never vanishes. This contradicts with (18). So subcase (b_1) is ruled out. It remains to settle subcase (b_2) that is $B(0) \neq 0$. Then $g(z_0) = 0$ for some z_0 implies $g''(z_0) = -C(0)/B(0) = d$; d a constant. If z_0 is a simple zero, then

$$K(z) = \frac{g''(z) - d}{g(z)} \tag{19}$$

is holomorphic at z_0. Now suppose that g vanishes at z_0 with multiplicity $p = 2$. If $p = 2$ it follows from (2) that g'' vanishes at z_0. Hence K is also holomorphic at z_0. If p is greater than 2, then (2) implies $C(0) = 0$. But this would result that g'' vanishes at z_0 with the same multiplicity as g, which is impossible. Hence K is an entire function and we are going to show that K, in fact, is a constant. Assume that K is not a constant. First of all, from (18) and (19), we have

$$\begin{aligned} T(r,K) &= m(r,K) \leq m(r,g''/g) + m(r,d/g) \\ &= S(r,g) \end{aligned} \tag{20}$$

Now the differential equation (2) for g can be expressed as

$$A(w)w'^2 + dB(w) + C(w) + K(z)wC(w) = 0 \tag{21}$$

Since K is not a constant, we conclude, by using Lemma 2, from (21) that A is a divisor of $dB + C$ and $wC(w)$, that is $wC(w) = A(w)A_1(w)$ and $dB(w) + C(w) = A(w)B_1(w)$, where A_1, B_1, are polynomials. Thus (21) becomes

$$w'^2 + B_1(w) + K(z)A_1(w) = 0 \tag{22}$$

It is easy to verify, according to Lemma 2, that $B_1(w) = b_0 + b_1w + b_2w^2$

and $A_1(w) = a_0 + a_1 w + a_2 w^2$, where $b_i (i = 1,2)$ and $a_i (i = 0,1,2)$ are constants. Differentiating equation (22) and substituting w'' by d + kw, we get

$$[2d + b_1 + a_1 K + w\,(2K + 2b_2 + 2b_2 K)]\, w' = -K'[a_0 + a_1 w + a_2 w^2]$$

Hence, according to Lemma 2,

$$w' = K_1 + K_2 w \tag{23}$$

where K_1 and K_2 are meromorphic functions satisfying: $T(r,K_i) = S(r,g)$; $i = 1,2$. Replacing w'^2 in (22) by $K_1^2 + 2K_1K_2w + K_2^2w^2$ and we get, by comparing coefficients, $K^2 + b_2 + a_2K = 0$. On the otherhand, it follows from (23) that $K_2 = \frac{-a_1K'^2}{2(a_2 + 1)\, K + 2a_2}$. The first result leads to $m(r,K) = 2m(r,K_2) + 0(1)$, while the second result yields $m(r,K_2) = S(r,K)$, if $a_2 = -1$, and $m(r,K_2) = m(r,K') + 0(1) \le m(r,K) + S(r,K)$ if $a_2 = -1$. We will get contradiction either case, unless K is a constant. But then the assertion follows as before. This completes the proof of Theorem 1.

3.2. PROOF OF THEOREM 2

Let $F(z) = f(g(z))$ and $F^{(n)}(z) = h(g(z))$. From this we have

$$f^{(n)}(g)D_n[g] + f^{(n-1)}(g)D_{n-1}[g] + \ldots\ldots + f'(g)D_1[g] = h(g) \tag{24}$$

where $D_n[g] = g'^n$, $D_{n-1}[g] = \frac{n(n-1)}{2} g'^{n-2}g''$, ..., $D_1[g] = g^{(n)}$. In general, $D_j[g]$ is a homogeneous differential polynomial in g of degree j (see [6]). Since $T(r,D_j[g]) = m(r,D_j[g]) \le jm(r,g) + S(r,g)$, Lemma 1 is applicable to equation (24). If $f^{(n)}$ not vanishes identically, then it follows from Lemma 1 that

$$D_n[g] = R_{1,n-1}(g)D_{n-1}[g] + \ldots. + R_{1,1}(g)D_1[g] + R_{1,0}(g) \tag{25}$$

where R_{1j} are rational functions. Thus we have

$$\sum_{j=1}^{n-1} [f^{(j)}(g) + R_{1,j}f^{(n)}(g)]\; D_j(g) = h(g) - f^{(n)}(g)R_{1,0}(g) \tag{26}$$

Applying Lemma 1 once more we get

$$D_{n-1}[g] = R_{2,n-2}(g)D_{n-2}[g] + \ldots. + R_{2,1}(g)D_1[g] + R_{2,0}(g) \tag{27}$$

and this gives

$$\sum_{j=1}^{n-2} [f^{(j)}(g) + (R_{1,j}(g) + R_{1,n-1}(g)R_{2,j}(g))\, f^{(n)}(g)$$

$$+ R_{2,j}(g)f^{(n-1)}(g)]\, D_j[g]$$

$$= h(g) - f^{(n)}(g)R_{1,0}(g) - f^{(n-2)}(g)R_{2,0}(g) \tag{28}$$

If the procedure does not break down, we finally arrive

$$D_n[g] = g'^n = R(g) \tag{29}$$

where R is a rational function. But the procedure can break down only if at some stage Lemma 1 is not applicable. But this means that one of the functions occurring in equations similar to (26) and (28) vanish identically, i.e.

$$A_n(w)f^{(n)}(w) + \dots\dots + A_{j+1}(w)f^{(j+1)}(w) + f^{(j)}(w) \equiv 0$$

To complete the proof of Theorem 2 we have to show that (29) has no other transcendental entire solutions then $c_1 + c_2e^{c_3z}$. First of all, it follows from Lemma 2 that R is a polynomial of degree at most n. By Wittich's Theorem [8] R has to be of degree n,

$$R(w) = c(w-t_1)^{p_1} \dots\dots (w-t_k)^{p_k} \tag{30}$$

where c is a constant, $t_i \neq t_j$ for $1 \leq i \leq j \leq k$, and $p_1 + p_2 + \dots + p_k = n$. If $k = 1$, we have $g'^n = c(g - t_1)^n$ or $g' = t(g - t_1)$, which gives $g(z) = t_1 + e^{tz+s}$. Now assume $k \geq 2$. Then it is easy to see that every t_j is a completely ramified value of g, therefore we must have $k = 2$. Let z_0 be a zero of $g-t_j$ of order k_j. Then $n(k_j - 1) = p_jk_j$ or $n(1 - \frac{1}{k_j}) = p_j$. This yields $n = p_1 + p_2 = n(1-\frac{1}{k_1}) + n(1 - \frac{1}{k_2}) = n(2 - \frac{1}{k_1} - \frac{1}{k_2})$ or $\frac{1}{k_1} + \frac{1}{k_2} = 1$. It follows that $k_1 = k_2 = \frac{1}{2}$, and hence $p_1 = p_2 = \frac{n}{2}$. Then (30) becomes

$$g'^n = c(g - t_1)^{\frac{n}{2}} (g - t_2)^{\frac{n}{2}}$$

or $g'^2 = c_0(g - t_1)(g - t_2)$, which leads to $g(z) = c_1 + c_2\cos(c_3z + c_4)$. This also completes the proof of Theorem 2.

REFERENCES

1. F. Gross, On factorization of meromorphic functions; *Trans. Amer. Math. Soc., 131* (1968), 215.

2. F. Gross and C. C. Yang, Further results on common right factors of a meromorphic function and its derivatives; *Yokohama Math. J.*, to appear.

3. W. K. Hayman, *Meromorphic Functions,* Oxford Press, 1964.

4. A. Z. Mokhouko and V. D. Mokhouko, Estimates for the Nevanlinna characteristics of some classes of meromorphic functions and their applications to differential equations; *Sib. Math. J., 15* (1974), 921-934.

5. V. Ngoom and I. V. Ostrovskii, The logarithmic derivative of a meromorphic function; *Akad. Nauk. Armjan SSR Doklady, 91* (1975), 272-277 (Russian).

6. N. Steinmetz, Über die fakoriserbaren Lösungen gewöhnlicher differentialgleichungen; *Math. Zeit., 170* (1980), 169-180.

7. N. Steinmetz, Bemerkung on einen Satz von Yosida, complex analysis; Proc. Colloq. Joensuu, Finland 1978, *Lect. Note Math., 747* (1979), 369-377.

8. H. Wittich, Ganze tranzendente Lösungen algebraischer differentialgleichungen; *Math. Ann., 122* (1950), 221-234.

ON REFINED DEFECT RELATIONS OF HOLOMORPHIC CURVES IN COMPLEX PROJECTIVE SPACE

Chen-Han Sung

Department of Mathematics
University of Notre Dame
Notre Dame, Indiana

1. INTRODUCTION

Let C be the complex line, and CP^m be the m-dimensional complex projective space. In terms of the homogeneous coordinates of CP^m, a *holomorphic curve*

$$f := (f_0 : \cdots : f_m) : C \to CP^m$$

can be considered as a system of entire functions $\{f_j\}_{j=0}^m$ without common zeros. In particular, a holomorphic curve $f : C \to CP^1$ is just a meromorphic function by identifying CP^1 with the Riemann sphere $C \cup \{\infty\}$.

The *order function* of $f = (f_0 : \cdots : f_m)$ is given by

$$\begin{aligned} T(r,f) &:= (1/2\pi) \int_0^{2\pi} \ell n\, [\Sigma_{j=0}^m |f_j(re^{i\theta})|^2]^{1/2}\, d\theta \\ &- (1/2\pi) \int_0^{2\pi} \ell n\, [\Sigma_{j=0}^m\, f_j(r_0 e^{i\theta})\ ^2]^{1/2}\, d\theta \end{aligned} \tag{1.1}$$

for $r \geq r_0$, where r_0 is some fixed positive constant. The *lower order* of f is defined by

$$\mu(f) := \liminf_{r \to +\infty} \ell n\, T(r,f)/\ell n\, r \tag{1.2}$$

For a hyperplane D in CP^m, we let $n(r,D,f)$ denote the number of points of intersection, counted with multiplicities, between D and $f(\{\xi \varepsilon C | |\xi| < \})$. The *counting function* is defined by

$$N(r,D,f) := \int_{r_0}^{r} n(u,D,f)\, du/u \tag{1.3}$$

for $r \geq r_0$. As

$$0 \leq N(r,D,f) \leq T(r,f) + O(1)$$

we let

$$\delta(D,f) := 1 - \limsup_{r \to +\infty} N(r,D,f)/T(r,f) \tag{1.4}$$

be the *defect* of D for f. Observe that $\delta(D,f) = 1$ whenever f(C) misses the hyperplane D in CP^m.

A holomorphic curve $f : C \to CP^m$ is said to be *linearly nondegenerate* if f(C) is not contained in any hyperplane in CP^m. The first major result in the value distribution theory of holomorphic curves is

THEOREM I. Let $\{D_\alpha\}_\alpha$ be a family of hyperplanes in general position in CP^m. Let $f : C \to CP^m$ be a linearly nondegenerate holomorphic curve in CP^m. Then

$$\Sigma_\alpha \delta(D_\alpha,f) \leq (m + 1) \tag{1.5}$$

Furthermore, this bound is sharp.

This was proved by H. Cartan [3] in 1933. In 1941, L. Ahlfors [1] reproved and extended Cartan's result to the "osculating curves" of f, which he cast in a geometric form.

For m = 1, those $\{D_\alpha\}_\alpha$ are distinct points $\{a_\alpha\}_\alpha$ on the Riemann sphere, and (1.5) reduces to the Nevanlinna defect relation [11]

$$\Sigma_\alpha \delta(a_\alpha,f) \leq 2 \tag{1.6}$$

for a meromorphic function f on C.

In this paper we are mainly concerned with the question of whether any more than the defect relation (1.5) can be said about the deficiencies of a holomorphic curve in the complex projective space.

We shall first study a problem on meromorphic functions along this direction in the following section. Next, we will consider its counterpart for holomorphic curves in CP^m. Finally, we will give a brief outline of the proof for our main result on holomorphic curves.

2. MEROMORPHIC FUNCTIONS

In 1939, O. Teichmüller [17, p. 167] suggested that, in addition to (1.6), certain conditions including finite order on the meromorphic function f might imply

$$\Sigma_\alpha \delta^{1/2}(a_\alpha, f) < +\infty \tag{2.1}$$

It was W. H. J. Fuchs [5] who established (2.1) in 1957 under only the assumption that f be of finite lower order. His work was subsequently refined by V. Petrenko [12], and I. Kazakova and I. Ostrovskii [9] in 1964. They concentrated primarily on the bounds for the sum in (2.1). In 1965, A. Edrei [4, p. 85] gave an alternative proof of Fuchs' Theorem.

By modifying a construction technique of A. A. Gol'dberg [61], [7], W. Hayman [8], in 1964, produced examples of meromorphic functions of finite order for which $\Sigma_\alpha \delta^\nu(a_\alpha, f)$ diverges for every $\nu < 1/3$. He also observed [8, p. 98] that the convergence of

$$\Sigma_\alpha \delta^{1/3}(a_\alpha, f)$$

may be made arbitrarily slow in the sense that, given any convergent positive series $\Sigma_\alpha C_\alpha$, there exists a meromorphic function of finite lower order such that $\delta^{1/3}(a_\alpha, f)$ may be greater than a constant multiple of C_α for $\alpha = 1$ to $+\infty$.

Above all, Hayman [8, p. 90] showed that if f has finite lower order, then

$$\Sigma_\alpha \delta^\nu(a_\alpha, f) < +\infty \tag{2.2}$$

for every $\nu > 1/3$. This indicates that if (2.2) remains true for $\nu = 1/3$, the resulting theorem would be the best possible.

Following Hayman's approach, V. Petrenko [13] proved

$$\Sigma_\alpha \delta^{1/3}(a_\alpha, f) \cdot [\log e/\delta(a_\alpha, f)]^{-1} < +\infty$$

in 1966. A year later, E. Bombieri and P. Ragnedda [2] obtained

$$\Sigma_\alpha \delta^{1/3}(a_\alpha, f) \cdot [\sigma(\delta(a_\alpha, f))]^{-1} < +\infty$$

for suitable functions $\sigma(t)$ satisfying $\int \sigma(t)\, dt/t < +\infty$. Finally,

A. Weitsman [18], in 1972, proved that if $\mu(f) < +\infty$, then

$$\Sigma_\alpha \delta^{1/3}(a_\alpha,f) < +\infty \tag{2.3}$$

Like Fuchs [5] and Hayman [8] in their earlier work, Weitsman studied the behavior of the derivative f' at places where the meromorphic function f is close to deficient values.

The necessity of finite lower order for (2.3) can be seen by examples due to Fuchs and Hayman [8, p. 80].

It is not too hard to see that (2.3) remains true for meromorphic functions, with finite lower order, over a compact Riemann surface with a finite number of points deleted. However, it is not known whether an analogous result holds for meromorphic functions over an arbitrary parabolic Riemann surface.

Some progress has been made recently. It suggests that the answer to this question probably is "yes".

THEOREM II. ([16]) Let f be a meromorphic function over a parabolic Riemann surface, which is defined by an algebroid function with linearly independent coefficients. If f is of finite lower order with $a_1, a_2, \cdots$ as its set of Nevanlinna deficient values, then

$$\Sigma_\alpha \delta^{1/3}(a_\alpha,f) < +\infty \tag{2.4}$$

Further, the exponent 1/3 cannot be decreased.

For a meromorphic function over C, this is (2.3).

3. HOLOMORPHIC CURVES

We next discuss some generalizations of Hayman's (2.2) and Weitsman's (2.3) to holomorphic curves in CP^m, and we begin with some definitions.

A holomorphic curve $f : C \to CP^m$ is said to be *d-nondegenerate* if f(C) is not contained in any hypersurface of degree less than or equal to d in CP^m. Note that 1-nondegenerate is just *linearly nondegenerate*, while ∞-nondegenerate means *algebraically nondegenerate*.

A family of hypersurfaces $\{D_\alpha\}_\alpha$ of degree d in CP^m will be said to have *normal crossings* if the corresponding hyperplanes, under a Veronese mapping (cf. [14])

$$\begin{aligned} \eta_d : \quad & CP^m \quad \to \quad CP^n \\ (z_0:\cdots:z_m) & \to (\cdots: z_0^{j_0}\cdots z_m^{j_m} :\cdots) \end{aligned} \tag{3.1}$$

where $j_0 + \cdots + j_m = d$ and $n := \binom{m+d}{d} - 1$, are in general position.

In 1974, V. I. Krutin' proved (cf. MR52, #8503) the convergence of $\Sigma_\alpha \delta^\nu(D_\alpha,f)$ for $\nu > 1/3$ in [10] under the assumption that $f : C \to CP^m$ is a linearly nondegenerate holomorphic curve and $\{D_\alpha\}_\alpha$ is an "admissible" family of hyperplanes in CP^m. This generalizes (2.2).

In 1979, (2.3) was extended in [15] from the case of meromorphic functions to the case of holomorphic curves by the following result.

THEOREM III. ([15]) Let $\{D_\alpha\}_\alpha$ be a family of hypersurfaces of degree d having normal crossings in CP^m. Let $f : C \to CP^m$ be a holomorphic curve of finite lower order. If f is d-nondegenerate, then

$$\Sigma_\alpha \delta^{1/3}(D_\alpha,f) < +\infty \tag{3.2}$$

Further, for each $m \geq 1$, the exponent 1/3 cannot be decreased.

It seems likely that the nondegeneracy condition of f for (3.2) can be reduced to $f \not\equiv$ constant, but we have not been able to show this.

Our study centers around those places where the curve f approaches some deficient hypersurface and the quantity

$$\omega(f) := |W(f)/||f||^{m+1}| \tag{3.3}$$

converges to zero, where

$$W(f) := W(f_0 : \cdots : f_m) := \begin{vmatrix} f_0 & \cdots & f_m \\ f_0' & \cdots & f_m' \\ \vdots & & \vdots \\ f_0^{(m)} & \cdots & f_m^{(m)} \end{vmatrix} \tag{3.4}$$

and

$$||f||^2 := \Sigma_{j=0}^m |f_j|^2 \tag{3.5}$$

Note that $|\omega(f)|$ is, when m = 1, the usual "spherical derivative" of the meromorphic function $f = (f_0, f_1)$.

Our argument applies without essential changes to the case that f is a holomorphic curve over a compact Riemann surface with a finite number of points deleted. Thus our Theorem III holds in this case.

To see the sharpness of the exponent 1/3 in (3.2), we have the following.

THEOREM IV. ([15]) Suppose that $\{\eta_\alpha\}_{\alpha=1}^{\infty}$ is a sequence of positive numbers such that $\Sigma_\alpha \eta_\alpha = 1$. Let $\{D_\alpha\}_{\alpha=1}^{\infty}$ be an arbitrary family of hyperplanes in general position in CP^m. Then there exists a linearly nondegenerate transcendental holomorphic curve $f : C \to CP^m$ of order 1 such that

$$\delta(D_\alpha, f) > \frac{2}{9(m+1)} \cdot \eta_\alpha^3 \qquad (3.6)$$

for all but at most m(m + 1) of the α's.

By choosing $\eta_\alpha = C \cdot \alpha^{-1} \cdot \{\ell n(\alpha + 2)\}^{-2}$, where C is a suitable constant, we deduce that $\Sigma_\alpha \delta^\nu(D_\alpha, f)$ may diverge for every $\nu < 1/3$. More generally we see that $\{\delta^{1/3}(D_\alpha, f)\}$ can be taken termwise greater than some multiple of an arbitrary positive sequence in ℓ^1.

The idea of the construction in Theorem IV goes back to the classical case m = 1, due essentially to Gol'dberg [6] and refined by Hayman [8].

4. PROOF OF THEOREM III

Observe from (3.1) that $\eta_d \circ f$ is a 1-nondegenerate holomorphic curve in CP^n with $\mu(\eta_d \circ f) < +\infty$, if the holomorphic curve f is d-nondegenerate in CP^m with $\mu(f) < +\infty$. Therefore, it is enough to show the following.

THEOREM V. ([14]) Let $\{D_\alpha\}_\alpha$ be a family of hyperplanes in general position in CP^m. Let $f : C \to CP^m$ be a linearly nondegenerate transcendental holomorphic curve in CP^m, with $\mu(f) < +\infty$. Then there is a constant $K = K(\mu, m; \delta(D_j, f), j = 1, \cdots, m+1) < +\infty$ such that

$$\Sigma_{\alpha=1}^{N} \delta^{1/3}(D_\alpha, f) < K \qquad (4.1)$$

for every $N > 0$. Further, for each $m \geq 1$, the exponent 1/3 cannot be decreased.

An outline of the proof for Theorem V will be given below, with details to appear in [14].

Fix N. We may assume that $N > 2m + 1$ and that $\{\alpha \mid \delta(D_\alpha, f) > 0\}$ has at least $2(m + 1)$ elements, say $\alpha = 1, 2, \cdots, 2(m+1), \cdots$.

For each hyperplane

$$D_\alpha := \{Z = (z_0 : \cdots : z_m) \in CP^m \mid \Sigma_{j=0}^m a_\alpha^j z_j = 0\}$$

we define

$$F_\alpha := \Sigma_{j=0}^m a_\alpha^j z_j$$

where $A_\alpha := (a_\alpha^0 : \cdots : a_\alpha^m)$ is a unit vector in C^{m+1}. Since no $(m + 1)$ of the A_α's are linearly dependent,

$$W(f) = c(\alpha_0, \cdots, \alpha_m) \cdot W(F_{\alpha_0}, \cdots, F_{\alpha_m}) \tag{4.2}$$

for some nonzero constant $c(\alpha_0, \cdots, \alpha_m)$, where W is the Wronskian defined in (3.3). Let

$$w(F_{\alpha_0}, \cdots, F_{\alpha_m}) := W(F_{\alpha_0}, \cdots, F_{\alpha_m}) / \prod_{j=0}^m F_{\alpha_j} \tag{4.3}$$

Note that $W(f)$ and $w(F_{\alpha_0}, \cdots, F_{\alpha_m})$ are meromorphic functions but $\omega(f)$ is not.

Following Weitsman [18], we study the behavior of the functions $\omega(f)$ and $w(F_{\alpha_0}, \cdots, F_{\alpha_m})$ in annuli around the Pólya peaks $\{r_n\}$ of order μ of $T(r,f)$, where $\mu = \mu(f) < +\infty$. For our investigation, the relevant property of Pólya peaks is that $r_n \to \infty$ and for any fixed $\sigma > 1$,

$$T(r,f) \leq (r/r_n)^\mu \cdot T(r_n, f) \cdot (1 + o(1)) \tag{4.4}$$

uniformly in $r_n \leq r \leq \sigma r_n$. We assume these intervals $[r_n, \sigma r_n]$ pairwise disjoint.

Using the lemma on the logarithmic derivative, we have

LEMMA 1. If $\sigma > 1$ is fixed, then

$$m(r, W(f) / \prod_{j=0}^m F_{\alpha_j}) = o(T(r,f))$$

$$T(r, w(F_{\alpha_0}, \cdots, F_{\alpha_m})) \leq K(m) \cdot T(r,f) \cdot (1 + o(1))$$

as $r \to \infty$ through the intervals $r_n \leq r \leq \sigma r_n$.

From the property of Pólya peaks (4.4) and the inequality [11]

$$(1/r) \int_1^r \ell n^+ M(t,g)\, dt \leq K \cdot T(2r,g)$$

we conclude

LEMMA 2. There exist sequences $\{r_n'\}$ and $\{r_n''\}$ satisfying

$$r_n \leq r_n' \leq 2r_n, \qquad 5r_n \leq r_n'' \leq 6r_n$$

and such that

$$\ell n^+ M(r_n', 1/\omega(f)) \leq K \cdot T(r_n,f)$$

$$\ell n^+ M(r_n'', 1/\omega(f)) \leq K \cdot T(r_n,f)$$

where $K = K(\mu)$ and $n > n_0$.

Next we introduce a sequence

$$\beta_n := \tau \cdot \{[T(r_n,f)]^{1/2} + \ell n\, 2\} \tag{4.5}$$

$n = 1, 2, \cdots$, where $\tau \geq 1$ is some constant to be determined below. We then define

$$E_n := \{\xi = re^{i\theta} \mid r_n \leq r \leq 6r_n;\ \ell n|\omega(f(\xi))| < -(m+2)\beta_n\}$$

$$F_n(\Lambda) := \{\xi = re^{i\theta} \mid r_n \leq r \leq 6r_n;$$

$$\ell n|F_\alpha(\xi)/||f(\xi)||\,| < -\beta_n, \text{ all } \alpha \in \Lambda\}$$

$$G_n(\Lambda) := F_n(\Lambda) \setminus \bigcup_{\Sigma \supsetneq \Lambda} F_n(\Sigma)$$

$E_n(\Lambda)$:= the set formed by the components of E_n which have a nonempty intersection with $G_n(\Lambda)$

$E_n(\Lambda,r)$:= the argument set in $[0,2\pi]$ of $E_n(\Lambda) \cap \{|\xi| = r\}$

$E_n(\alpha,r)$:= the argument set in $[0,2\pi]$ of $E_n \cap F_n(\{\alpha\}) \cap \{|\xi| = r\}$, where $\Lambda \subset \Sigma \subset \{1, 2, \cdots, N\}$ and $1 \leq |\Lambda| < |\Sigma| \leq m$.

It follows from the above and Lemma 1 that for each α, we have

$$\delta(D_\alpha, f) \cdot T(r,f) \cdot (1+o(1)) \leq (1/2\pi) \int_{E_n(\alpha,r)} \ln|\omega(f(re^{i\theta}))|^{-1}\, d\theta \tag{4.6}$$

for $r \to \infty$ with $r_n \leq r \leq 6r_n$.

By consideration of (3.3), (4.2), (4.3) and the length of the level curves of $|w(F_{\alpha_0}, \cdots, F_{\alpha_m})|$ in $|\xi| < 6r_n$, we claim that

LEMMA 3. There exist some $\tau \geq 1$ for (4.5) and some $n_1 > 0$ such that $\{E_n(\Lambda,r)\}_{n,\Lambda}$ are pairwise disjoint for $n > n_1$ and $r_n \leq r \leq 6r_n$.

Then, by a known lemma of Edrei and Fuchs together with (4.4) and (4.6), we obtain

LEMMA 4. There exists $\theta_0 = \theta_0(\delta(D_j,f),\ j = 1,\cdots,m+1) > 0$ such that for each $\alpha > m + 1$,

$$0 \leq \limsup_{\substack{r\to\infty \\ r_n \leq r \leq 6r_n}} \Theta_n(\alpha,r) \leq 2\pi - \Theta_0$$

where $\Theta_n(\alpha,r)$ denotes the angular measure of $E_n(a,r)$.

Now, let $p_{n,\alpha}$ be the number of zeros of $\omega(f)$ in

$$H_n(\alpha) := F_n(\{\alpha\}) \cap \{r_n' < |\xi| < r_n''\} \cap E_n$$

where $\{r_n'\}$ and $\{r_n''\}$ are given by Lemma 2.

Applying Lemma 2, Lemma 4, and the estimates on harmonic measures in [18] to some harmonic measures on the components of $H_n(\alpha)$, we are led to

$$\int_{E_n(\alpha,r)} \ln|\omega(f(re^{i\theta}))^{-1}\, d\theta$$

$$\leq 2\pi\beta_n + K \cdot p_{n,\alpha}\left[\int_{r_n}^{6r_n} \theta_n(\alpha,t)\, dt/t\right]^2$$

$$+ K \cdot T(r_n,f) \cdot \exp\left[-K\int_{r_n'}^{r} dt/t \cdot \Theta_n(\alpha,t)\right]$$

$$+ K \cdot T(r_n,f) \cdot \exp\left[-K\int_{r}^{r_n''} dt/t \cdot \theta_n(\alpha,t)\right] \tag{4.7}$$

for $r_n' < r < r_n''$ and $\alpha > m + 1$.

Combining (4.6), (4.7) and $(a + b + c)^{1/3} \leq (a^{1/3} + b^{1/3} + c^{1/3})$, we get

$$\delta^{1/3}(D_\alpha,f) \leq K \cdot \left[p_{n,\alpha}/T(r_n,f) + \int_{r_n}^{6r_n} \Theta_n(\alpha,t)\, dt/t\right]$$

for sufficiently large n, where $K = K(\mu, m; \delta(D_j,f), j = 1,\cdots,m+1)$ and $\alpha > m + 1$.

By (4.4), Lemma 1 and Lemma 3, we have (with n being fixed)

$$\Sigma_{\alpha=m+2}^{N} \int_{r_n}^{6r_n} \Theta_n(\alpha,t)\, dt/t \leq 2\pi \cdot \ln 6 \cdot m$$

and

$$\Sigma_{\alpha=m+2}^{N} p_{n,\alpha} \leq K(\mu,m) \cdot T(r_n,f) \cdot (1 + o(1))$$

Therefore,

$$\Sigma_{\alpha=1}^{N} \delta^{1/3}(D_\alpha,f) \leq K(\mu, m; \delta(D_j,f), j = 1,\cdots,m+1) < +\infty$$

for every $N > 0$ and so Theorem V follows.

5. REMARKS

Recently, people have used the defect relation of meromorphic functions to study the factorization theory. For example, see R. Goldstein's paper in J. London Math. Soc. in 1970, and several papers of M. Ozawa in Kodai Math. Sem. Rep. during 1968. Many of their results have dealt with entire functions of finite order, which have maximal deficiency sum.

In order to extend the factorization theory to holomorphic mappings from C^n into CP^m, we have to find out what is the proper form for a holomorphic mapping to be a prime factor. Even in the case n = 1 and m = 2, we do not know. We hope that the ideas and techniques being used here could give us some clue and results for such a factorization theory in the future.

REFERENCES

1. L. V. Ahlfors, The theory of meromorphic curves; *Acta Soc. Sci. Fenn., Ser. A 3,* No. 4 (1941), 3-31.

2. E. Bombieri and P. Regnedda, Sulle deficienze delle funzioni meromorfe di ordine inferiore finite; *Rend. Sem. Fac. Sci. Univ. Cagliari, 37* (1967), 23-38.

3. H. Cartan, Sur les zeros des combinaisons lineaires de p fonctions holomorphes donnees; *Mathematica, Cluj, 7* (1933), 5-31.

4. A. Edrei, Sums of deficiencies of meromorphic functions; *J. Analyse Math., 14* (1965), 79-107.

5. W. H. J. Fuchs, A theorem on the Nevanlinna deficiencies of meromorphic functions of finite order; *Ann. of Math., 68* (1958), 203-209.

6. A. A. Gol'dberg, On the deficiencies of meromorphic functions; *Dokl. Akad. Nauk SSSR, 98* (1954), 893-895.

7. A. A. Gol'dberg, On the set of deficient values of meromorphic functions of finite order; *Ukrain. Mat. Zh., 11* (1959).

8. W. Hayman, *Meromorphic Functions,* Oxford, 1964.

9. I. Kazakova and I. Ostrovskii, On the defects of meromorphic functions of small orders; *Zapiski Mekh.-Matem. Fak. Khar'k. Gos. Un. i Khar'k. Matem. Obsc., 30* (1964), 70-73.

10. V. I. Krutin', A remark on the growth and distribution of values of p-dimensional entire curves; *Vestnik Har'kov Gos. Univ. No. 113 Mat. i Meh. Vyp., 39* (1974), 31-36, 122. (MR 52 #8503).

11. R. Nevanlinna, *Le théorème de Picard-Borel et la théorie des fonctions meromorphes*, Gauthier-Villars, 1929.

12. V. Petrenko, Defects of meromorphic functions, *Dokl. Akad. Nauk SSSR, 158* (1964), 1030-1033.

13. V. Petrenko, Some estimates for the magnitudes of defects of a meromorphic function; *Sibirsk. Mat. Zh., 7* (1966), 1319-1336.

14. C. H. Sung, On the deficiencies of holomorphic curves in complex projective space, to appear.

15. C. H. Sung, A theorem of Nevanlinna deficiencies for meromorphic functions over parabolic Riemann surfaces; Presented at the Conference in Classical Complex Analysis, Purdue University, March, 1980.

16. C. H. Sung, Two theorems of Nevanlinna deficiencies for meromorphic functions over parabolic Riemann surfaces, to appear.

17. O. Teichmüller, Vermutungen und Sätze über die Wertverteilung gebrochener Funktionen endlichen Ordnung; *Deutsche Math., 4* (1939), 163-190.

18. A. Weitsman, A theorem on Nevanlinna deficiencies; *Acta Math., 128* (1972), 41-52.

SOME RESULTS ON THE UNIQUENESS OF THE FACTORIZATION OF CERTAIN, PERIODIC, ENTIRE FUNCTIONS

Hironobu Urabe

Kyoto University of Education
Fushimi-ku, Kyoto
Japan

INTRODUCTION

An entire function F(z) is said to have a factorization with left-factor f(z) and right-factor g(z), when, under

$$F(z) = f(g(z)) = f \circ g(z)$$

f is meromorphic and g is entire (now g may be meromorphic when f is rational). F(z) is called to be *prime (pseudo-prime)*, if every factorization of the above form implies that f or g is linear (rational, resp.). When factors are restricted to entire functions, the factorization is called to be in entire sense (F may be prime in entire sense, etc.).

Now recall a question of Gross [8] which asks whether there exists an entire, periodic function which is prime. Concerning this question, Ozawa [13] succeeded in proving the existence of such a function of finite order and Baker-Yang [3] exhibited such one in the case of order infinity (also cf. Gross-Yang [9]). In the previous paper [17], the present author has shown further that there exist periodic entire functions which together with their (all) derivatives are prime.

In this note, we'll consider certain composite, periodic, entire functions and prove that they are "uniquely factorizable in entire sense (see Theorems 1, 2 and 3). Now, assume that a non-constant entire function $F(z)$ has two factorizations;

$$F(z) = f_1 \circ f_2 \circ \cdots \circ f_n(z) = g_1 \circ g_2 \circ \cdots \circ g_m(z)$$

into non-linear entire factors. Here, if $m = n$ and if with suitable linear polynomials $T_j(z)$ $(j = 1, \cdots, n-1)$ the following relations

$$f_1(z) = g_1 \circ T_1^{-1}(z), \quad f_2(z) = T_1 \circ g_2 \circ T_2^{-1}(z), \quad \cdots\cdot, \quad f_n(z) = T_{n-1} \circ g_n(z)$$

hold identically, then the two factorizations are called *equivalent* (in entire sense). If any two factorizations into non-linear, prime, entire factors are equivalent, we say that $F(z)$ is *uniquely factorizable*. Of course, prime functions may be considered to be uniquely factorizable.

Note that, so far, several results have been obtained on the uniqueness of the factorization for certain entire functions (cf. [13], [10] etc.).

Well, at the final section, we shall state one more result (Theorem 4) which is of some different nature but closely related to the "unique-factorizability."

For an entire function $f(z)$, we shall denote by $\rho(f)$ $(\underline{\rho}(f))$ the order (the lower order, resp.) of f and by $\rho^*(f)$ the exponent of convergence of the zeros of $f(z)$. For example,

$$\rho(f) = \limsup_{r \to \infty} \frac{\log \log M(r, f)}{\log r} = \limsup_{r \to \infty} \frac{\log T(r, f)}{\log r}$$

where $M(r, f)$ is the maximum modulus of $f(z)$ for $|z| = r$ and $T(r, f)$ is the Nevanlinna characteristic function of $f(z)$ ($\underline{\rho}(f)$ is given by lim inf). Further $\rho^*(f)$ is known to be equal to the order of $N(r, 1/f)$ (cf. [10] p. 16-25).

1. STATEMENT OF RESULTS

THEOREM 1. Let m be a non-negative integer. Let $P(z)$, $Q(z)$, $R(z)$ ($\not\equiv$ const.) and $h(z)$ (trans. and $h(0) \neq 0$) be entire functions. Assume that there exists an integer N ($\geqq 1$) such that

$$M(r, P) \leq \exp[(\log r)^N]$$

for all sufficiently large values of r, and that the order of $h(e^z)$ is finite and h has at least one simple zero or two zeros with co-prime multiplicities. Further assume that if $\rho(h(e^z)) = 1$, Q and R are both polynomials and if $1 < \rho(h(e^z)) < \infty$,

$$\rho[Q(e^z) + R(e^{-z})] < \rho[h(e^z)].$$

Then setting

$$F(z) = (z \cdot e^{P(z)}) \circ (h(e^z) \cdot \exp[mz + Q(e^z) + R(e^{-z})]),$$

F(z) is uniquely factorizable.

When m is a negative integer, instead of Theorem 1, we can prove, for instance, the following results.

THEOREM 2. Let m be a negative integer. Assume that K(z) is a polynomial which has at least one simple zero or two zeros with co-prime multiplicities, $K(0) \neq 0$ and that $\deg K < |m|$. If P(z) and Q(z) are two non-constant polynomials, then the function

$$F(z) = (z \cdot e^{P(z)}) \circ (K(e^z) \cdot \exp[mz + Q(e^z)])$$

is uniquely factorizable.

THEOREM 3. Let m be a negative integer. Assume that K(z) is a polynomial which has at least one simple zero or two zeros with co-prime multiplicities, $K(0) \neq 0$ and that $\deg K > |m|$. If P(z) and R(z) are two non-constant polynomials, then the function

$$F(z) = (z \cdot e^{P(z)}) \circ (K(e^z) \cdot \exp[mz + R(e^{-z})])$$

is uniquely factorizable.

REMARKS. It will follow, from Theorem 2, that the function

$$F(z) = (ze^z) \circ ((e^z - 1)\exp[-2z + e^z])$$

is uniquely factorizable (cf. [14] p. 348) and, from Theorem 3, that the function

$$F(z) = (ze^z) \circ ((2e^{2z} - e^z + 1)\exp[-z + e^{-z}])$$

is uniquely factorizable. Note that the right factors of F(z) above are not prime, however prime in entire sense:

$$(e^z - 1)\exp[-2z + e^z] = \left(\frac{w-1}{w^2}\cdot e^w\right)\circ(e^z)$$

and

$$(2e^{2z} - e^z + 1)\exp[-z + e^{-z}] = \left(\frac{w^2 - w + 2}{w}\cdot e^w\right)\circ(e^{-z})$$

2. LEMMAS WHICH WILL BE NEEDED LATER

LEMMA 1. (Pólya [15]). Suppose f(z), g(z) and h(z) are nonconstant entire functions such that f(z) = g(h(z)). If h(0) = 0, then there exists a constant c with $0 < c < 1$ such that

$$M(cM(r/2, h), g) \leqq M(r, f) \quad (r \geqq r_o).$$

Here h(0) = 0 is not essential, which means that if we add the condition $r \geqq r_o$ (for sufficiently large values of r), then the condition h(0) = 0 can be removed. Note that the inequality $M(r, f) \leqq M(M(r, h), g)$ is valid, by the maximum modulus principle.

LEMMA 2. (Edrei-Fuchs [4]). Let f(z) and g(z) be two transcendental entire functions. If $\rho^*(f)$ is positive, then $\rho^*(f(g(z))$ is infinite.

LEMMA 3. (cf. [1], [7]). Let f(z) ($\not\equiv$ const.) and g(z) be two entire functions such that the order $\rho(f)$ is less than 1/2. If f(g(z)) is periodic, then g(z) must be so.

LEMMA 4. (Baker-Gross [2]). Let F(z) be entire and periodic mod h(z) with period b, where h(z) is a non-constant entire function of order less than one. (This means that $F(z + b) - F(z) \equiv h(z)$.). Then every right factor g(z) of F(z) is of the form:

$$g(z) = H_1(z) + h_1(z)\cdot\exp[H_2(z) + cz]$$

where $H_j(z)$ (j = 1, 2) are periodic entire functions with the same peiod b, c is a constant and $h_1(z)$ ($\not\equiv 0$) is an entire function of order less than 1. If h(z) is a polynomial, then $h_1(z)$ is also a polynomial.

REMARK. For the validity of Lemma 4, it is sufficient to assume that h(z) (of order less than one) does not vanish identically. That is, the condition that h is nonconstant is unnecessary. (See the proof of Theorem 1 in [2]). Hence we can assume in Lemma 4 that F(z) has the form:

$$F(z) = H(z) + h(z),$$

where H(z) is periodic with period b and h(z) is nonconstant and of order less than 1.

LEMMA 5. (Kobayashi [11]). Let f(z) be an entire function such that

$$\liminf_{r \to \infty} T(r, f)/r = 0$$

Assume that there exists an unbounded sequence $\{w_n\}$ such that all the roots of the equations

$$f(z) = w_n \quad (n = 1, 2, \cdots)$$

lie in a half plane (Re z $\geqq$ c for some real number c, say), then f(z) is a polynomial of degree at most two.

LEMMA 6. (Yang-Urabe [18]). Let f(z) be an entire function of finite order and of positive lower order. Suppose that $\alpha(z)$ is an entire function of order not greater than 1/2 and $\alpha \notin \underline{H}$, i.e.

$$\liminf_{r \to \infty} \frac{\log \log M(r, \alpha)}{\log \log r} = \infty \qquad (*)$$

Then $\rho^*(\alpha(f(z))$ is infinite.

LEMMA 7. (cf. [18]). Let f(z) be an entire function of finite order. Then if $\alpha \in \underline{H}$(i.e. the left hand side of (*) in the above Lemma 6 is finite), the lower order $\underline{\rho}(\alpha(f))$ is finite.

LEMMA 8. (cf. [18]). Let f(z) be an entire function. If the lower order $\underline{\rho}(e^{f(z)})$ is finite, then f(z) is a polynomial.

LEMMA 9. (Goldstein [15]). Let F(z) be an entire function of finite order such that $\delta(a, F) = 1$ for some $a \neq \infty$, where $\delta(\ , F)$ denotes the Nevanlinna deficiency. Then F(z) is pseudo-prime.

3. PROOF OF THEOREM 1

Let

$$F(z) = f(g(z)) = f\circ g(z) \tag{1}$$

with nonlinear entire functions f and g. We must prove that there exists a linear polynomial T(z) such that the relations

$$f(z) = (ze^{P(z)})\circ T(z)$$

and

$$g(z) = T^{-1}(z)\circ(h(e^z)\cdot\exp[mz + Q(e^z) + R(e^{-z})])$$

hold identically.

Since all the zeros of $h(e^z)$ are contained in a half plane (which is checked from the fact that h is entire), it will be clear that g(z) cannot be a polynomial of degree $\geqq 3$. Well, if g is a polynomial of degree 2, then, setting

$$g(z) = a(z - z_o)^2 + b$$

for some constants a ($\neq 0$), b and z_o, and substituting the variable w = $z - z_o$, F must satisfy the identity

$$F(-w + z_o) = F(w + z_o) \tag{2}$$

From (2), we conclude that

$$h(e^{-w+z_o})/h(e^{w+z_o}) \neq 0 \tag{3}$$

However, this is clearly impossible. In fact, because h(z) has an infinite number of zeros (note that h is entire, transcendental and of order zero), hence the numerator of the left hand side of (3) have zeros whose real parts are smaller than any given value, while the denominator have zeros whose real parts are greater than any given value. Thus (3) cannot hold. Hence g must be transcendental.

Also f must be transcendental. Here assume that f would be a polynomial. Then, if f has at least two distinct zeros, noting $\rho(g) = \infty$ in this case and using Borel-Nevanlinna's theorem (cf. [12] p. 72), we have $\rho^*(f(g)) = \infty$, which shows $\rho^*(F) = \rho^*(h(e^z)) = \infty$. Hence $\rho(h(e^z)) = \infty$ also. This is contrary to the assumption. If f has only one zero, its multiplicity is not less than 2 (say $k \geqq 2$), since f is nonlinear. Then the multiplicities of the zeros of $F(z) = f(g(z))$ are all some multiples of k.

This is again contrary to the assumption concerning h(z). Here note that the function e^z has no multiple values.

Since f and g under (1) are both transcendental (proved as above), from Lemma 2, we have $\rho^*(f) = 0$ so that we can write

$$f(z) = \alpha(z)e^{A(z)} \tag{4}$$

where $\alpha(\neq \text{const.})$ and A are entire functions with $\rho(\alpha) = 0$.

Subsequently we'll treat the following three cases separately:

(i) α has just one zero-point,

(ii) α has at least two and a finite number of zeros,

(iii) α has an infinite number of zeros (hence α is transcendental).

The case (i). We may put

$$f(z) = z \cdot e^{A(z)} \tag{5}$$

and

$$g(z) = h(e^z) \cdot e^{B(z)} \tag{6}$$

Here A and B are nonconstant entire functions. Note that B(z) is nonconstant, because R(z) is so (cf. (7) below).

Then from (1), we obtain the following identity:

$$A(h(e^z) \cdot e^{B(z)}) + B(z) = mz + Q(e^z) + R(e^{-z})$$

$$+ P\{h(e^z) \cdot \exp[mz + Q(e^z) + R(e^{-z})]\} \tag{7}$$

We wish to note that either if

$$\rho(h(e^z)) = 1 \tag{8}$$

B(z) is of exponential type, or if

$$1 < \rho(h(e^z)) < \infty \tag{9}$$

then

$$\rho(B(z)) < \rho(h(e^z)) \tag{10}$$

In fact: Obviously (for sufficiently large values of r)

$$M(r, F) \leqq (e^{e^{(\log r)^N}}) \circ (e^{e^{Kr}}) \leqq e_3(K'r)$$

or

$$M(r, F) \leqq (e^{e^{(\log r)^N}}) \circ (e^{e^{r^P}}) \leqq e_3(Nr^P)$$

according as (8) or (9), for some constants K and K' (= NK) and for some number p with $\rho[Q(e^z) + R(e^{-z})] < p < \rho(h(e^z))$ (in the case (9)). Here $e_n(r)$ denotes the n-times iterated function of e^r.
While, in view of Lemma 1, we have

$$M(r, f(g(z))) \geqq M(cM(r/2, g), f) \geqq e^{dcM(r/2, g)}$$

Here we use the fact that $M(r, f) \geqq e^{dr}$ $(r \geqq r_0)$ for some positive constant d. Further, as we may assume that B(z) is transcendental (otherwise, nothing is to be proved), and hence the lower order of $e^{B(z)}$ is infinite (by Lemma 8), we know $T(r, g) = (1 + o(1))T(r, e^B)$, and so we obtain

$$M(r/2, g) \geqq [M(r/4, e^B)]^{c'} \geqq [M(cM(r/8, B), e^z)]^{c'}$$
$$= [e^{cM(r/8, B)}]^{c'} = e^{cc'M(r/8, B)} \quad (r \geqq r_0)$$

for some positive constant c' (we may take c' = 1/6). Here we've used the inequality $\log M(r, g) \geqq T(r, g) \geqq 1/3 \cdot \log M(r/2, g)$ (cf. [10] p. 18) and Lemma 1. Hence from the above three inequalities, we deduce (noting F = f(g)), for some c",

$$e_2[c''M(r/8, B)] \leqq e_3(K'r) \quad (\text{or} \leqq e_3(Nr^p))$$

so that

$$M(r, B) \leqq 1/c'' \cdot e^{8K'r} \quad (\text{or} \leqq 1/c'' \cdot e^{8Nr^p})$$

hold for sufficiently large values of 4, which shows that B(z) is of exponential type or that (10) is valid.

From (7), we have the following identity

$$A(h(e^z) \cdot e^{B(z + 2\pi i)}) - A(h(e^z) \cdot e^{B(z)})$$
$$= 2m\pi i - [B(z + 2\pi i) - B(z)] \tag{11}$$

Now the zeros (including multiplicities) of

$$e^{B(z + 2\pi i)} - e^{B(z)} = e^{B(z)}[e^{B(z + 2\pi i) - B(z)} - 1] \tag{12}$$

must be contained to the zeros of the right hand side of (11). Noting this fact and that B(z) is of finite order, we conclude from (11) and (12) that

$$B(z + 2\pi i) - B(z) = \text{const.} = c \text{ (say)} \tag{13}$$

From (11) and (13), by noting $B(z + 2n\pi i) - B(z) = cn$, we'll obtain

$$A(e^{cn}h(e^z)e^{B(z)}) - A(h(e^z)e^{B(z)})$$
$$= 2mn\pi i - cn = (2m\pi i - c)n \tag{14}$$

for any integer n. Well, the left hand side of (14) remains bounded with respect to n for the positive or negative integers n, according as $|e^c| \leqq 1$ or $|e^c| \geqq 1$. Hence, we have

$$c = 2m\pi i \tag{15}$$

where m is the nonnegative integer given in the Theorem. Then from (13), we can write

$$B(z) = mz + H(z) \tag{16}$$

where H is an entire function such that $H(z + 2\pi i) = H(z)$. In this case, using (16), the identity (7) can be written as

$$A[h(e^z)\cdot e^{mz + H(z)}] = Q(e^z) + R(e^{-z}) - H(z)$$
$$+ P\{h(e^z)\cdot \exp[mz + Q(e^z) + R(e^{-z})]\} \tag{17}$$

Here we consider the set (including multiplicities)

$$E = \{ z \,;\; h(e^z) = 0\} \tag{18}$$

Taking, for z, those points belonging to the set E, from (17), we have

$$Q(e^z) + R(e^{-z}) - H(z) = A(0) - P(0) \qquad (z \in E) \tag{19}$$

We wish to show that the left hand side of (19) must be constant. If $\rho(h(e^z)) = 1$, noting (16), the left hand side of (19) is of exponential type (in this case B(z) is of exponential type, as is shown before, and that Q and R are both polynomials, by assumption), while the function $H(e^z)$ is (or order 1 and) of maximal type, since h is transcendental. Therefore it follows that (19) is valid only when the left hand side of it is constant itself. Next, if $1 < \rho(h(e^z)) < \infty$ noting (10) and (16), the order of the left hand side of (19) is less than that of $h(e^z)$. By the fact as we'll check it in the following: $\rho^*(h(e^z)) = \rho(h(e^z))$, and the above reasoning, we have a contradiction from (19) unless the left hand side of (19) is constant. Here we'll check the assertion mentioned above. Now assume

$$\rho^*(h(e^z)) \quad < \quad \rho(h(e^z)) \tag{20}$$

We can write

$$h(e^z) = \Pi(z)\cdot e^{L(z)} \tag{20'}$$

where $\Pi(z)$ is the canonical product which is constructed by the zeros of $h(e^z)$ and $L(z)$ is a polynomial (in this case, $e^{L(z)}$ is of finite order). If (20) is valid, then

$$\rho(\Pi(z)) = \rho^*(h(e^z)) < \rho(h(e^z)) = \rho(e^{L(z)})$$

(by a well-known theorem due to Borel). But in this case,

$$\delta(0, \quad \Pi(z)e^{L(z)}) = 1$$

Then, by Lemma 9, the function $\Pi(z)e^{L(z)}$ must be pseudo-prime. Since $h(z)$ and e^z are both transcendental, we have a contradiction under the representation (20'). Hence (20) is not valid.

Thus, we have that the following identical relation holds;

$$H(z) = Q(e^z) + R(e^{-z}) + c'$$

where c' is a constant. Hence, from (16)

$$B(z) = mz + Q(e^z) + R(e^{-z}) + c' \tag{21}$$

and from (7) or (17), we have $A(e^{c'}z) = P(z) - c'$ and so

$$A(z) = P(e^{-c'}z) - c' \tag{22}$$

Hence, by (5), (6), (21) and (22), we have

$$f(z) = e^{-c'}\cdot z\cdot e^{P(e^{-c'}z)} = (z\cdot e^{P(z)})\circ(e^{-c'}z)$$

and

$$g(z) = h(e^z)\cdot\exp[mz + Q(e^z) + R(e^{-z}) + c']$$

$$= (e^{c'}z)\circ(h(e^z)\cdot\exp[mz + Q(e^z) + R(e^{-z})])$$

Putting

$$T(z) = e^{-c'}z \tag{23}$$

we obtain

$$f(z) = (ze^{P(z)})\circ T(z)$$

and

$$g(z) = T^{-1}(z)\circ(h(e^z)\cdot\exp[mz + Q(e^z) + R(e^{-z})])$$

Thus the factorizations of F(z) given and $f\circ g(z)$ are equivalent.

The case (ii). We may put

$$f(z) = S(z)e^{A(z)} \tag{24}$$

where A(z) is an entire function and S(z) is a polynomial with degree not less than 2 such that S(z) has at least two distinct zeros. In this case, since $\rho(S(g(z)) = \rho^*(S(g(z)) = \rho^*(h(e^z)) < \infty$ (cf. the argument used in p.8),

$$\rho(S(g)) = \rho(g) < \infty \tag{25}$$

Hence

$$S(g(z)) = h(e^z)\cdot e^{U(z)} \tag{26}$$

for some polynomial U(z) (by $\rho(h(e^z)) < \infty$ and (25)), and so that by (1) we have the following identity:

$$A(g(z)) = mz - U(z) + Q(e^z) + R(e^{-z})$$

$$+ P\{h(e^z)\cdot\exp[mz + Q(e^z) + R(e^{-z})]\} \tag{27}$$

In the following, we shall deduce a contradiction, by using the above relations (26) and (27).

If

$$mz - U(z) = \text{const.} + c \text{ (say)} \tag{28}$$

then, by applying Lemma 3 to (26), we know that g(z) is periodic and hence we may put

$$g(z) = p(e^{z/k}) + q(e^{-z/k}) \tag{29}$$

for some entire functions p(z) and q(z), where k is a natural number. From (26), (28) and (29), we have

$$S\{p(e^{z/k}) + q(e^{-z/k})\} = h(e^z)\cdot e^{mz - c}$$

so that, replacing the variable ($e^{z/k}$ to z), we obtain the identity;

$$S\{p(z) + q(1/z)\} = h(z^k)\cdot e^{-c}\cdot z^{mk}$$

Here S(z) is a nonconstant polynomial, h is entire and m is a nonnegative integer (by assumption) and that k is a natural number. Therefore, it follows that q(z) must be constant. Further, the identity (27), substituting the variable as before, can be written as

$$A\{p(z) + c'\} = c + Q(z^k) + R(1/z^k)$$

$$+ P\{h(z^k)\cdot z^{mk}\cdot \exp[Q(z^k) + R(1/z^k)]\} \tag{30}$$

Well the left hand side of (30) is entire, while the right hand side is not so (note that R(z) is nonconstant and so, for instance, the real part of $R(1/z^k)$ can tend to $-\infty$ as z tends to zero suitably). This is a contradiction. Hence (28) is not valid.

Next assume

$$mz - U(z) \neq \text{const.} \tag{31}$$

In this case, since U(z) is a polynomial, by applying Lemma 4 to (27), we get

$$g(z) = H_1(z) + P_1(z)\cdot \exp[cz + H_2(z)]$$

where $H_j(z + 2\pi i) = H_j(z)$ (j = 1, 2) are entire, $P_1(z)$ is a polynomial and c is a constant. Note that $P_1(z)$ is nonconstant, which is easily seen by comparing the growth of the both sides of (27) on the upper (or lower) part of the imaginary axis (cf. the argument below).
Further $H_2(z)$ must be constant. Indeed, under

$$g(z + 2\pi i) - g(z) = \{P_1(z + 2\pi i)e^{2\pi ic} - P_1(z)\}\cdot \exp[cz + H_2(z)]$$

we have $\rho[g(z + 2\pi i) - g(z)] < \infty$ also from (25). Hence, by Lemma 8, it follows that $H_2(z)$ (periodic) is constant. Here note that the function $\{P_1(z + 2\pi i)e^{2\pi ic} - P_1(z)\}$ cannot vanish identically, since $P_1(z)$ is a nonconstant polynomial. Thus we may write

$$g(z) = H_1(z) + P_1(z)\cdot e^{cz}$$

The relations (26) and (27) can be written as

$$S\{H_1(z) + P_1(z)\cdot e^{cz}\} = h(e^z)\cdot e^{U(z)} \tag{32}$$

and

$$A\{H_1(z) + P_1(z)\cdot e^{cz}\} = mz - U(z) + H_3(z) \tag{33}$$

where $H_3(z + 2\pi i) = H_3(z)$, entire, is the periodic part of the right hand side of (27). Now it must be that

$$c \in R \text{ (the set of real numbers)} \tag{34}$$

Indeed, if $c \notin R$, then the left hand side of (33) remains bounded along the upper or lower part of the imaginary axis (since $P_1(z)\cdot e^{cz}$ tends to zero as z tends to ∞ there), while the right hand side is unbounded (and has polynomial-like-growth) there. This contradiction shows that the assertion (34) is valid.

However, the fact (34) implies that, as z tends to ∞ along the imaginary axis, the left hand side of (32) is unbounded and of polynomial growth, while the right hand side (whose essential part is $e^{U(z)}$ in this situation) either remains bounded or (otherwise) has at least order-one-like-growth. This is a contradiction. Thus (31) is not valid either.

The case (iii). In this case, we have

$$f(z) = \alpha(z)\cdot e^{A(z)} \tag{35}$$

where α and A are entire such that α is transcendental and $\rho(\alpha) = 0$.

Noting again that the zeros of $h(e^z)$ are all contained in a half-plane and that g is transcendental, in view of Lemma 5, we know at first

$$\rho(g) \geqq 1$$

Next, note that $\rho(g) < \infty$ also in this case, since $\rho^*(h(e^z)) < \infty$, and by applying Lemma 6, we know

$$\alpha \in \underline{H}$$

so that by Lemma 7 we have

$$\underline{\rho}(\alpha(g)) < \infty$$

Hence we have the identity

$$\alpha(g(z)) = h(e^z)\cdot e^{V(z)} \tag{36}$$

where $V(z)$ is a polynomial in view of Lemma 8.

If

$$mz - V(z) = \text{const.} \tag{37}$$

then applying again Lemma 3 to the relation (36), we see that $g(z)$ is periodic so that $g(z)$ can be written as in (29). By noting that m is a nonnegative integer and h is entire, as before, we conclude that $q(z)$ under

(29) is constant. Further, by using the similar relation as (30), we obtain a contradiction. Hence (37) is not valid.

If

$$mz - V(z) \neq \text{const.} \tag{38}$$

by using (almost) the same argument such as in the case (ii) (around (32) and (33)), also we can deduce a contradiction. In this case, however, we shall need the minimum modulus theorem concerning the entire functions of order less than 1/2. Note that $\rho(\alpha) = 0$.

Thus the case (iii) does not occur.

Consequently, only the case (i) is possible, which shows that F(z) is uniquely factorizable. Thus we have done.

We note here that, if we use the arguments used in p. 12 and p. 17, we can show the following fact:

PROPOSITION. Let h(z) be a nonconstant entire function with $\rho(h) = 0$ such that $h(0) \neq 0$. Then we have necessarily

$$\rho^*(h(e^z)) = \rho(h(e^z))$$

We can prove the above proposition, by applying Lemmas 5, 6 and 9. Indeed, if $\rho^*(h(e^z)) < \infty$, then by using Lemmas 5 and 6, we can show that $\rho(h(e^z)) < \infty$ necessarily. Hence, by using Lemma 9, we can conclude the assertion of the proposition (cf. the argument in p. 12). Note that in the above argument, it will be sufficient to consider the case where h is transcendental.

4. NOTES ON THE PROOFS OF THEOREMS 2 AND 3

The proof of Theorem 2 or 3 is a little easier than that of Theorem 1. Hence we'll remark only about the several points to be noted.

Now we assume that F(z) given in Theorem 1 or 2 has a factorization such as (1) with nonlinear entire functions f and g.

At first, we can show that f and g both are transcendental. The fact that g cannot be a (nonlinear) polynomial follows by showing that the identity (2) does not hold. This deduction can be done very elementarily, using the assumptions that K is a polynomial and that Q (or R) is a nonconstant polynomial. Indeed, if F satisfies (2), then substituting the variable e^z by z, we can rewrite (2) as follows;

$$\begin{aligned}
&\exp\{P(K(e^{z_0}\cdot 1/z)\cdot z^{-m}\cdot e^{mz_0}\cdot \exp[Q(e^{z_0}\cdot 1/z) + R(e^{-z_0}\cdot z)])\\
&- P(K(e^{z_0}\cdot z)\cdot z^{m}\, e^{mz_0}\cdot \exp[Q(e^{z_0}\cdot z) + R(e^{-z_0}\cdot 1/z)])\\
&+ Q(e^{z_0}\cdot 1/z) - R(e^{-z_0}\cdot 1/z)\}\\
&= \frac{K(e^{z_0}\cdot z)}{K(e^{z_0}\cdot 1/z)}\cdot z^{2m}\cdot \exp[Q(e^{z_0}\cdot z) - R(e^{-z_0}\cdot z)]
\end{aligned} \tag{39}$$

In the above relation, only one of Q and R is not constant. Under the assumption in Theorem 2 or 3, the left hand side of (39) must have the point $z = 0$ as the essential singular point (this is so concerning the part $\{\cdots\}$ in the above), while the right hand side has (at most) a pole at $z = 0$. This is a contradiction.

Next, we note about the argument to conclude that an identical relation such as (21) holds. Under the situation around (17), letting

$$H(z) = s(e^{z/k}) + t(e^{-z/k})$$

for some natural number k, where $s(z)$ and $t(z)$ are certain polynomials (note that $B(z)$ and hence $H(z)$ is of exponential type), and noting $K(z)$ is also a polynomial, we can proceed as follows:
Substituting the variable $e^{z/k}$ by z, the relation (17) reduces to

$$\begin{aligned}
&A(K(z^k)\cdot z^{km}\cdot e^{s(z) + t(1/z)}) = -s(z) - t(1/z) + Q(z^k) + R(1/z^k)\\
&+ P(K(z^k)\cdot z^{km}\cdot e^{Q(z^k) + R(1/z^k)})
\end{aligned} \tag{40}$$

By tending z to infinity along a certain half-line issued from the origin, we see at first that $\deg s(z) = \deg Q(z^k)$ and the leading coefficients of $s(z)$ and $Q(z^k)$ are equal (cf. the argument around p. 107-108 in [16]). Next, as we may do so, taking a half-line on which $e^{s(z)}$ and $e^{Q(z^k)}$ both tend to zero, we can conclude that

$$s(z) - Q(z^k) = \text{const.}$$

since a polynomial $s(z) - Q(z^k)$ becomes bounded there (under (40)). Further tending z to the origin along a suitable path (such that $1/z$ tends to infinity along a suitable half-line), similarly we have

$$t(z) - R(z^k) = \text{const.}$$

Hence we obtain $H(z) = Q(e^z) + R(e^{-z}) + c'$, and so, from (17) we see that (22) is valid. Thus in the case (i), we have the equivalent factorizations.

About the cases (ii) and (iii): In these cases, under the relation such as (27), we shall only need to remark how to exclude the case corresponding to (28) (Note that in the case such as (31) the treat is essentially same as in the proof of Theorem 1). We consider the relation such as (26). Then under the case such as (28), (29) is valid. Here, again by substituting the variable $e^{z/k}$ to z, we get an identical relation such as

$$A(p(z) + q(1/z)) = \text{const.} + Q(z^k) + R(1/z^k)$$

$$+ P\{K(z^k)\cdot z^{mk}\cdot \exp[Q(z^k) + R(1/z^k)]\} \tag{30'}$$

When

$$Q(z) \neq \text{const.} \quad \text{and} \quad R(z) = \text{const.} \tag{41}$$

and

$$\deg K < |m| \tag{42}$$

as is assumed in Theorem 2, from (26) (h is replaced by K) we have that

$$p(z) = \text{const.} \tag{43}$$

since in this case the relation (26) becomes

$$S(p(z) + q(1/z)) = K(z^k)\cdot z^{mk} \tag{44}$$

whose right hand side tends to zero as z tends to infinity (by (42)) and hence it is holomorphic at the point $z = \infty$. From (43), the relation (30') can be written as

$$A(q(1/z) + \text{const.}) = \text{const.} + O(z^k) + P(K(z^k)\cdot z^{mk}\cdot e^{Q(z^k)}) \tag{45}$$

But, this identity cannot hold, because the right hand side of (45) has $z = \infty$ as the essential singular point, while the left hand side is holomorphic there.

Next, when

$$Q(z) = \text{const.} \quad \text{and} \quad R(z) \neq \text{const.} \tag{46}$$

and

$$\deg K > |m| \tag{47}$$

as is assumed in Theorem 3, then the right hand side of (30') has $z = 0$ as the essential singular point. By this fact and by noting that q(z) is a polynomial, we see that $A(z)$ must be transcendental. Hence we know $p(z) = \text{const.}$ Indeed, if p is nonconstant, then the left hand side of (30') has $z = \infty$ as an essential singular point, while the right hand side is at most a pole there, a contradiction. However if

$$p(z) = \text{const.}$$

then (26) becomes

$$S(q(1/z) + \text{const.}) = K(z^k)\cdot z^{mk} \tag{48}$$

By (47), if we tend z to ∞, the right hand side of (48) tends to ∞, while the left hand side tends to a finite value (since q and S are entire). This is a contradiction. Thus we can rule out the case (28).

The above deduction can be applied to the case (iii) without essential changes (cf. the argument in the proof of Theorem 1).

Finally, we note the following result

THEOREM 4. Let

$$F(z) = (P(z)\cdot e^{Q(z)})\circ(h(e^z)) \equiv P(h(e^z))\cdot\exp[Q(h(e^z))]$$

where P, Q and h are nonconstant entire functions which satisfy the conditions

$$\rho(P(h(e^z))) < \infty, \quad \underline{\rho}(Q(h(e^z)) < \infty$$

and further that the function i.e. $P(h(z))$, $P(h(o)) \neq 0$ has at least one simple zero or two zeros with coprime multiplicities. Now assume that $F(z)$ is factorized such as

$$F(z) = f\circ g(z) \tag{*}$$

where f and g are two nonlinear entire functions.

Then we have necessarily

$$f(z) = S(z)\cdot e^{A(z)}$$

and

$$g(z) = r(e^{z/k}) = r(z)\circ(e^{z/k})$$

such that the following identical relations hold;

$$S(r(z)) = e^c\cdot P(h(z^k))$$

and

$$A(r(z)) = Q(h(z^k)) - c$$

where c is a constant, k is a natural number and that S(z), A(z) and r(z) are certain nonconstant entire functions.

In particular, the function g(z) in (*) must be periodic.

The proof of Theorem 4 can be done by using the similar argument as in the proof of Theorem 1. But, as the result seems to be of some interest, hence we'll give its detailed proof elsewhere.

REFERENCES

1. I. N. Baker, On some results of A. Renyi and C. Renyi concerning periodic entire functions; *Acta Sci. Math. (Szeged), 27* (1966), 197-200.

2. I. N. Baker and F. Gross, Further results on factorization of entire functions; *Proc. Symp. Pure Math. (Amer. Math. Soc.), 11* (1968), 30-35.

3. I. N. Baker and C. -C. Yang, An infinite order periodic entire function which is prime; *Lecture Notes in Math.* (Springer) *599* (Complex Analysis, Kentucky, 1976), 7-10.

4. A. Edrei and W. H. J. Fuchs, On the zeros of f(g(z)) where f and g are entire functions; *J. d'Analyse Math., 12* (1964), 243-255.

5. R. Goldstein, On factorization of certain entire functions; *J. London Math. Soc. (2), 2* (1970), 221-224.

6. R. Goldstein, On factorization of certain entire functions II; *Proc. London Math. Soc. (3), 22* (1971), 438-506.

7. F. Gross, On factorization of meromorphic functions; *Trans. Amer. Math. Soc., 131* (1968), 215-222.

8. F. Gross, Factorizations of entire functions which are periodic mod g; *Indian J. Pure and Appl. Math., 2* (1971), 561-571.

9. F. Gross and C. -C. Yang, On prime periodic entire functions; *Math. Zeit., 174,* (1980), 43-48.

10. W. K. Hayman, *Meromorphic Functions,* (Clarendon Press, Oxford, 1964).

11. T. Kobayashi, Distribution of values of entire functions of lower order less than one; *Kodai Math. Sem. Rep., 28* (1976), 33-37.

12. R. Nevanlinna, *Le théoréme de Picard-Borel et la théorie des fonctions méromorphes,* (Gauthier Villars, Paris, 1929).

13. M. Ozawa, Factorization of entire functions; *Tohoku Math. J., 27* (1975), 321-336.

14. M. Ozawa, On uniquely factorizable meromorphic functions; *Kodai Math. J., 1* (1978), 339-353.

15. G. Polya, On an integral function of an integral function; *J. London Math. Soc., 1* (1926), 12-15.

16. H. Urabe, Uniqueness of the factorization under composition of certain entire functions; *J. Math. Kyoto Univ., 18* (1978), 95-120.

17. H. Urabe, Infinite order, periodic, entire functions which together with their derivatives are prime; *Bull. Kyoto Univ. of Education, Ser. B, No. 55* (1979), 21-32.

18. C. -C. Yang and H. Urabe, On permutability of certain entire functions; *J. London Math. Soc. (2), 14* (1976), 153-159.

PROGRESS IN FACTORIZATION THEORY OF ENTIRE AND MEROMORPHIC FUNCTIONS

Chung-Chun Yang

Naval Research Laboratory
Washington, D.C.

INTRODUCTION

The theory of factorization of meromorphic functions is basically a study of the ways in which a given meromorphic function can be expressed as a composition of other meromorphic functions. More specifically, given a meromorphic function F, we are interested in whether or not F can be represented as the form $F = f_1 of_2 of_3 .. of_n$, where f_i are meromorphic functions. If none of the f_i's are linear we call the factorization reduced. If F(z) is representable as $f_1of_2o...of_n$ and $g_1og_2o...og_n$ and if with suitable linear transformations $_j$, j = 1, 2, ... , n-1

$$f_1 = g_1o\lambda_1, \; f_2 = \lambda_2^{-1}og_2o\lambda_2, \; \dots \; , \; f_n = \lambda_{n-1}^{-1}og_n$$

hold, then the two representations or factorizations are called to be equivalent. Particularly, a meromorphic function F(z) = fog(z) (=f(g)) is said to have f and g as left and right factors respectively provided that f is meromorphic and g is entire (g may be meromorphic when f is rational). F(z) is said to be prime (pseudo-prime) if every factorization of the above form into factors implies either f is linear or g is linear (either f is rational or g is a polynomial). Further, F is said to be left-prime (right-prime) if every factorization $F = f \circ g$ in the entire sense (i.e. only entire factors are considered) implies that f is linear whenever g is transcendental (g is linear whenever f is transcendental). Whenever only entire factors are considered the term E-prime, E-pseudo-prime will be used accordingly.

After the classical works of G. Julia [23,24] and P. Fatou [6] on the iteration and composition theory for polynomials or rational functions, Ritt [48] has investigated and obtained a fairly complete factorization theory for polynomials. An analogous theory for rational functions is still an unsettled one. Later on, I. N. Baker [1] studied the iteration and composition theory in the case of transcendental entire functions and obtained many results. In particular, he generalized the minimum modulus theorem concerning entire functions of order less than 1/2 ([2], Theorem 3) and further proved, using Fatou's theory of iteration, some interesting theorems concerning the permutability of two factors for a certain class of transcendental entire functions. In 1952, P. Rosenbloom [49] in the investigation of the fixed points of entire function stated without proof, that the function $F(z) = z+e^z$ is prime. This function has several special properties such as it has no fixed point, no multiple zeros, and it is periodic mod a nonconstant polynomial, and its derivatives has restricted zeros. Therefore, the primeness of e^z+z can be proved by several different approaches, and the proofs have suggested several directions in the development of the factorization theory. The primeness of e^z+z was subsequently proved by F. Gross [14]. Shortly after this Gol'dberg and Prokopovich [8] obtained a generalization of this result by using a different argument. Later on Gross and Baker [4] showed that for any nonzero polynomial p(z) and any nonconstant entire function $\alpha(z)$ of order less than one, the function it may be extremely difficult, if not impossible, to determine whether entire prime functions can indeed be formed.

Most functions which have been studied thus far are related to this primitive type of functions. It seems that given an arbitrary entire function it may be extremely difficult, if not impossible to determine whether or not it is prime.

The factorization theory has been studied by numerous authors in various aspects and many results have been obtained. However, no systematic research has actually been developed except for an isolated paper by Ritt [48] dealing with the special case of polynomials. Most of the results so far derived concern the impossibility of factorization, that is, the primeness, the pseudo-primeness of certain classes of entire and meromorphic functions, and uniqueness of the factorization of certain entire functions. The tools that have been employed in the study of factorization theory are mainly based on Nevanlinna's value distribution theory. Most classes of functions, which have been studied, are concerned with the following one or several factors: (1) the growth of the function, (2) roots distribution, (3) Taylor coefficients, (4) periodicity, (5) fixed-points, (6) the existence of defect values.

In this survey we attempt to collect results on factorization of meromorphic and entire functions obtained recently. We shall also recall some unsolved conjectures connected with the subject of prime functions.

1. PRIME FUNCTIONS WHICH ARE PERIODIC MOD h

According to [4], an entire function $F(z)$ is said to be periodic mod h (where h is an entire function) with period σ if $F(z+\sigma)-F(z) = h(z)$.

LEMMA 1.1. [4] Every right factor $g(z)$ of an entire function $F(z)$ which is periodic mod h with period σ, is of the form $g(z) = H_1(z)+h_1(z)e^{H_2(z)+az}$, where $H_i(z)$, $i = 1, 2$ are periodic entire functions with the same period σ, a is a constant, and $h_1(z)$ is an entire function of order less than or equal to one, if h is a polynomial, so is h_1.

The above lemma leads to

THEOREM 1.1. [4] If $p(z)$ is any nonconstant polynomial and let $a(\neq 0)$ and b be any two constants, then $F(z) = e^{az+b}+p(z)$ is prime.

The above result is obviously not true when $p(z)$ is a constant. This leads to an interesting question raised in [17, problem 3]: Does there exist a periodic entire function which is prime?

This question was answered affirmatively by Ozawa [34] for finite order (order 1) case and by Baker and Yang [5] for infinite order (hyperorder 1) case. Recently, Ozawa [39] constructed such functions for arbitrary finite order ρ $(1 \leqq \rho < \infty)$ as well as for finite hyperorder μ $(1 \leqq \mu < \infty)$. More recently, Urabe [55] exhibited a class of periodic entire functions of arbitrary growth which together with all their derivatives are prime.

It was indicated in [18] that Theorem 1.1 can be generalized as follows:

THEOREM 1.2. Let $h_1(z)$ be a nonzero polynomial and $h_2(z)$ ($\neq$ constant) be an entire function of order less than the degree of the polynomial $p(z)$. Then the only possible factorization of $h_1(z)e^{p(z)} + h_2(z)$ is of the form $h_1(z)^{p(z)} + h_2(z) = f(Q(z))$, where $Q(z)$ is a polynomial of degree m no greater than the degree of $p(z)$.

This result was also proved by R. Goldstein [10] as a special case of the following theorem.

THEOREM 1.3. Let m be a given positive integer, $p_m(z)$ a nonconstant polynomial of degree m, ϕ_m ($\neq$ constant) ψ_m ($\neq$ constant) entire functions of order less than m. Write $F(z) = \phi_m(z) + \psi_m(z)\exp(p_m(z))$ and suppose that $F(z) = f(g(z))$ with $f(z)$, $g(z)$ nonlinear entire functions. Then $g(z)$ is a polynomial of degree k and $f(z)$ is of order p_f such that $kp_f = m$ and $k \leqq m$.

If $p(z) = z$, the following generalization of theorem 1.1 is obtained.

THEOREM 1.4. If q, h ($q \neq 0$, $h \neq$ constant) are entire functions of order less than one, then $q(z)e^z + h(z)$ is prime.

By using the method of [18] and some elementary facts of algebraic functions, Gross and Yang [19] obtained the following result which is slightly stronger than that of Theorem 1.3.

THEOREM 1.5. Let $p(z)$ be a nonconstant polynomial of degree m, $h(z)$ ($\neq 0$) and $k(z)$ ($\neq$ constant) be two entire functions of order less than m. Then $h(z)e^{p(z)} + k(z)$ is either prime or it can be factorized as:

$$h(z)e^{p(z)} + k(z) = f(L(z)) \tag{E}$$

where $L(z)$ is a nonlinear polynomial of degree n, and $f(z) = \mu(z)\exp[cL(z)]^d + \beta(z)$ an entire function with the following three relations being satisfied:

(i) $n|m$ (i.e. $\frac{m}{n}$ is an integer) ,

(ii) $h(z)e^{p(z)} \equiv \mu(L(z))\exp[cL(z)]^d$, where $d = \frac{m}{n}$ an integer, c is a constant, and $\mu(z)$ is an entire function of order less than d,

(iii) $k(z) = \beta(L(z))$ where β is an entire function of order less than d.

Remarks. (1) The example $\exp(kz^m) + 1 = (z^k + 1) \circ \exp(z^m)$, $m \geqq 1$, $k > 1$, shows that the restriction on $k \neq$ constant in the theorem is a necessary one.

(2) Suppose that equation (L) holds. Then it follows from the assertions (ii) and (iii) that he^p and k have a nonlinear polynomial $L(z)$ as their common right factor, and hence he^p and k are not relatively R-polynomial prime. (Two entire functions, F_1, F_2 are called relatively R-polynomial prime if the only possible common right factor of F_1 and F_2 are linear functions.)

(3) From assertion (i) it follows particularly that $n \leqq m$, i.e., the degree of L is no greater than the degree of p. From these remarks the following five results were obtained.

THEOREM 1.6 [19]. Let $p(z)$ be a nonconstant polynomial of degree m and h ($\neq 0$), k ($\neq$ constant) be two entire functions of order less than m. If k has no nonlinear polynomial factor whose degree divides m, then $he^p + k$ is prime.

THEOREM 1.7 [19]. Let p and h be as in the above theorem. Let k be any nonconstant entire function of order less than m and let h_1 be an entire function of order less than m and $q(z)$ be a polynomial of degree less than m. Suppose that p and q are relatively R-polynomial prime. Then

$$h(z)e^{p(z)} + h_1(z)e^{q(z)} + k \text{ is prime.}$$

If one notes that the only possible nonlinear polynomial right factor for a periodic entire function is a polynomial of degree two (see e.g. [47]) then one can also derive the following conclusion from previous two theorems.

THEOREM 1.8 [19]. Let p, h, and k be as in theorem 1.6. Suppose further that the degree of p is odd and that k is a nonconstant periodic function. Then $he^p + k$ is prime.

THEOREM 1.9 [19]. Given any entire periodic function of finite order ρ, one can always find an entire function such that H is prime.

Proof. We simply choose $h = k = H$ and p be any polynomial of odd degree m with $m > \rho$.

THEOREM 1.10 [19]. Let p, h, and k as in Theorem 1.8 assume further that the order of h is less than one. Then $he^p + k$ is prime if p and k are relatively R-polynomial prime.

Remark. The order restriction on h is necessary. If we allow h to be of order greater than or equal to one, then the above result may not be true. For instance, $e^z e^{z^2-z} + z^2$ is not prime (in this example, $h = e^z$, $p = z^2-z$, and $k = z^2$).

When h, p, k are polynomials E. Mues [28] resolved the class of entire functions: $h(z)e^{p(z)} + k(z)$ as solutions of differential equations of Riccati type. He, by means of some analytic number theorems proved that every entire solution of Riccati type with polynomial coefficients is pseudo-prime and also obtained the same conclusions (ii) and (iii) as in Theorem 1.5.

In a forthcoming paper [50], Steinmetz determined solutions of the functional equation $w(z) = f \circ g\ (z)$, where $w(z)$ is a solution of the Schwarzian differential equation $\{w,z\} = q(z)$; $q(z)$ entire. As to factorization for monomorphic solutions satisfying linear differential equations we refer the reader to another paper of Steinmetz's [51].

Almost at the same period, G. S. Prokopovich by using arguments different from that of Gross-Yang and Mues proved the following general results:

THEOREM 1.11 [46]. Let $F(z) = \sum_{j=1}^{k} Q_j(z)e^{p_j(z)}$, $F(z) \neq c + Q_2(z)e^{p_2(z)}$ (c: a constant). If F(z) has a nontrivial factorization $F(z) = f(g(z))$, then g(z) must be a polynomial and for all j (j = 1, 2, ..., k). $Q_j(z) = q_i(g(z))$, $p_j(z) = p_i(g(z))$, where p_i and q_i are polynomials.

THEOREM 1.12 [46]. Let g(z), p(z) be polynomials respectively, of degree k and q; $q = k\ell$, ℓ is an integer, f(z), $\phi_1(z)$ and $\phi_2(z)$ are entire functions ($\phi_1(z) \neq$ constant, $\phi_2(z) \neq 0$), where

$$T(r,\phi_i) = o\,\{T(r,f(g))\}\ (i = 1,\ 2)\ \text{as}\ r \to \infty$$

and

$$f(g(z)) = \phi_1(z) + \phi_2(z)e^{p(z)}$$

then

$$\phi_1(z) = \varphi_1(g(z)),\quad \phi_2(z)e^{p(z)} = \varphi_2(g(z))e^{p_1(g(z))}\quad ,$$

$$f(z) = \varphi_1(z) + \varphi_2(z)\, e^{p_1(z)}$$

where $\varphi_i(z)$ are entire functions $(i = 1, 2)$, $\deg p_1(z) = \ell$, and

$$T(r,\varphi_i) = o\{T(r,f)\} \text{ as } r \to \infty$$

Remark. Here it is assumed that $g(z)$ is a polynomial while in theorem 1.5 it shows that g must be a polynomial.

For some entire function $F(z)$ which is periodic mod h, where h is a function of order greater than or equal to one, we have the following result:

THEOREM 1.13 [57]. Let $H_1(z)$ and $H_2(z)$ be two periodic functions with the same period σ. Then every right factor $g(z)$ of the entire function $ze^{H_1(z)} + H_2(z)$ satisfies the following difference equation:

$$cg(z + 2\sigma) - g(z + \sigma) + (1-c)g(z) = 0$$

for some nonzero constant c. Furthermore, if $c \neq \frac{1}{2}$, then

$$g(z) = \omega_1(z)e^{2k\pi iz} + \omega_2 e^{(\log b)z}$$

otherwise

$$g(z) = \omega_1(z)e^{2m\pi iz} + z\omega_2(z)e^{2n\pi iz}$$

where $\omega_1(z)$, $\omega_2(z)$ are two periodic entire functions with period σ, k, m and n are integers and $b = (1-c)/c$.

THEOREM 1.14 [57]. Let $H(z)$ be a nonconstant periodic entire function with period 1. Then every right factor $g(z)$ and left factor $f(z)$ of the function $z + e^{H(z)}$ are of the form $g(z) = H_1(z) + \ell_1(z)$ and $f(z) = G_1(z) + \ell_2(z)$ respectively, where $H_1(z)$ is a periodic entire function with period 1 and $G_1(z)$ is also a periodic entire function, $\ell_i(z)$, $i = 1, 2$ are linear functions.

Several classes of prime functions relating to the form $z + e^H$ have been found.

THEOREM 1.15 [11]. $H(z) + az$ is prime, where H is entire and periodic and satisfies, for any positive ε, the condition $M_H(r) < \exp(\exp(\varepsilon r))$ for an infinite sequence of r, depending on ε and a is a nonzero constant.

In [4] Baker and Cross also proved the following result.

THEOREM 1.16. Let $F(z) = H(z) + z$, where $H(z)$ is a periodic entire function of finite lower order, then $F(z)$ is prime.

Later on Urabe [53] obtained the following generalization of Theorem 1.16.

THEOREM 1.17. Let $F(z) = H(z) + Q(z)$, where $H(z)$ ($\neq$ constant) is an entire function of finite lower order which is periodic with period $2\pi i$ (say) and $Q(z)$ is a nonconstant polynomial, then $F(z)$ is left-prime in entire sense [i.e. only entire factors are considered]. Further if $Q(z)$ has no quadratic right factor, then $F(z)$ is prime.

This also provides an affirmative answer to an earlier conjecture of Gross ([15], conjecture 2).

THEOREM 1.18. For any integer $n \geqq 1$, $e_n(z) + z$ is prime, where $e_n(z)$ denotes the n-th iterate of e^z.

The case $n = 2$ follows from the Theorem 1.15. The more general case for $n > 2$ was posed in [15] and proved by Ozawa [36].

THEOREM 1.19 [15]. $e^{H(z) + (2s\pi i/\tau)z} + az$ is prime, where $H(z)$ is periodic of period τ, entire and of exponential type, s is any integer and a is a nonzero constant.

The following conjecture was raised in [57] and an affirmative answer hasn't been found yet.

Conjecture. Let H be a nonconstant periodic entire function, then $z + e^{H(z)}$ is prime.

Thus we have seen the following three classes of entire functions have yielded numerous types of prime functions. $J_o(\sigma) = \{F(z) = e^{H(z)} + az \mid e^{H(z)}$ periodic entire function with period σ; a is a complex number $\neq 0\}$, $J(\sigma) = \{F(z) = H(z) + az \mid H(z)$ periodic entire function with period σ; a is a complex number $\neq 0\}$,
$L(\sigma) = \{F(z) = H_1(z) + ze^{H_2(z)} \mid H_1, e^{H_2}$ both are periodic entire functions with period $\sigma\}$. Clearly,

$$J_0 \subset J \subset L$$

Theorem 1.14 says that if $F \in J_0(\sigma)$ and if $F(z) = f(g(z))$ is nontrivial factorization then $g \in J(\sigma)$ and $f \in L(\sigma^*)$ for an appropriate value of σ^*. In a report by S. Koont [26] the following results have been obtained.

THEOREM 1.20. Let $F(z)$ be an entire function with a nontrivial factorization $F(z) = f(g(z))$, then

(A) if $F \in L(\sigma)$, then $g \in L(\sigma)$,
(B) if $F \in L(\sigma)$, then $g \in J(\sigma)$,
(C) if $F \in J(\sigma)$, then $f \in J(\sigma^*)$ for an appropriate value of σ^*.

THEOREM 1.21 [26]. If $F \in J(\sigma)$ and $F = f_n \circ f_{n-1} \circ \dots \circ f_1 (n \geqq 2)$ is a reduced factorization of F, then for $i = 1, 2, \dots, n$, $f_i \in J(\sigma_i)$ for an appropriate value of σ_i. In fact, if $F(z) = H(z) + a$ and $f_i(z) = R_i + b_i z$, we have

$$a = \prod_{i=1}^{n} b_i$$

and

$$\sigma_i = \begin{cases} \sigma & \text{if } i = 1 , \\ \left(\prod_{j=1}^{i-1} b_j\right)\sigma & \text{if } 1 < i \leqq n . \end{cases}$$

The actual values of $\{b_i\}_{i=1}^{n}$ and $\{\sigma_i\}_{i=1}^{n}$ are easily obtained from the proof of Theorem 1.20 Part C.

Corollary. If $F \in L(\sigma)$ and $F = f_n \circ \dots \circ f_1 (n \geqq 2)$ is a reduced factorization of F, then for $i = 1, \dots, n-1$ we have $f_i \in J(\sigma_i)$ for an appropriate value of σ_i.

The next result in [26] concerns the number of factors in a reduced factorization of $F \in J(\sigma)$. We let $e_n(z)$ be the n-th iterate of e^z.

THEOREM 1.22. Let $F \in J(\sigma)$ and $F = f_n \circ \dots \circ f_1$ be a reduced factorization of F. If for every $\varepsilon > 0$,

$$M(r,F) < e_m(\varepsilon r) \quad \dots\dots\dots\dots \quad (*)$$

for a sequence of r's going to infinity, then $n \leqq m-1$.

We note the following two remarks made in [26].

Remark 1. Theorem 1.22 is sharp in the sense that there exist functions in $J(\sigma)$ with reduced factorizations containing $(m-1)$ factors, which satisfy the growth condition (*). For example, let $h(z) = e^z + z$ and $h_{m-1}(z) =$ $(m-1)$st iterate of $h(z)$. Then, clearly, $h_{m-1} \in J(2\pi i)$ has $m-1$ factors in a reduced factorization, and for every $\varepsilon > 0$

$M(r,h_{m-1}) < e_m(\varepsilon r)$

for all sufficiently large r.

Remark 2. In general, there is no upper bound on the number of factors in a reduced factorization, determined by the order of growth of the factorized function. For example, for $F(z) = e^z$, of order 1, there exist reduced factorizations of arbitrarily many factors. In fact given $n \geqq 2$, let $f_1(z) = e^{z/2^{n-1}}$ and $f_2(z) = \ldots = f_n(z) = z^2$. Then $f_n \circ f_{n-1} \circ \ldots \circ f_1$ is a reduced factorization of e^z consisting of n factors.

Finally, a corollary of Theorem 1.22 concerning factorization in $L(\sigma)$.

Corollary. Let $F \in L(\sigma)$ and $F = f_n \circ \ldots \circ f_1$ be a reduced factorization of F. If for every $\varepsilon > 0$

$M(r,F) < e_m(\varepsilon r)$

for a sequence of r's tending to infinity, then $n \leqq m$.

Remark. Results related to Theorem 1.20 were also obtained by Urabe [52] by a different argument. There the conclusion $f \in L(\sigma^*)$ was obtained in part A or B and that $g \in J(\sigma)$ was obtained in part C. The following three results were also derived in [52].

THEOREM 1.23. Let $F(z) = H(z) + cz \in J(\sigma)$ which is of finite lower order, then F(z) is prime.

THEOREM 1.24. Let $F(z) = H(z) + ze^z \in L(\sigma)$, then F(z) is prime.

THEOREM 1.25. Let $F(z) = ze^{H(z)} \in L(\sigma)$, then F is prime.

2. PRIME FUNCTIONS WHICH HAVE SOME RESTRICTION ON THE ZEROS OF CERTAIN VALUES

THEOREM 2.1 [33,I]. If F(z) is an entire function of finite order and it has a finite Picard exceptional values, then F(z) is E-pseudo-prime.

A more general and useful result is the following:

THEOREM 2.2 [9]. Let F(z) be an entire function of finite order suppose that $\delta(a,F) = 1$ for some $a \neq \infty$. Then F is E-pseudo-prime.

THEOREM 2.3 [33,I]. Let $F(z)$ be an entire function of finite order and it admits two perfect branched values then F is E-pseudo-prime. Here a perfect branched value ω of F means that $F-\omega$ has a finite number of simple zeros and has an infinite number of multiple zeros.

THEOREM 2.4 [33,I]. Let F be an entire function of finite order and there are p disjoint continuous curves Γ_j which extend to infinity and on which all the zeros of F lie and along which F is bounded. Then F is E-pseudo-prime.

As extensions of the previous results on functions of finite order the following are connected with certain special types of entire functions of infinite order.

THEOREM 2.5 [33,II]. Assume that $F(z)$ has the form $p(z)e^{H(z)}$ where $p(z)$ is a nonconstant polynomial and $H(z)$ an entire function of order less than one. Then $F(z)$ is E-pseudo-prime.

THEOREM 2.6 [33,II]. Assume that $F(z)$ has the form $p(z)\exp(\exp(z))$ where $p(z)$ is a nonconstant polynomial. Then $F(z)$ is E-pseudo-prime.

Remark. The above two results are also true if the assumption on F is replaced by F'. For instance $F(z) = \int_0^z e^{e^z}\,dz$ is E-pseudo-prime.

THEOREM 2.7 [33,III]. Let $F(z)$ be an entire function of finite order for which $F(z) = A$ for some $A(\neq \infty)$ has only real roots. Then $F(z)$ is E-pseudo-prime.

Examples. $z^p \sin^q z$ (p, q integers, $q \geqq 1$; $p \geqq -q$); $\frac{1}{\Gamma(z)}$, the nth Bessel function $J_n(z)$; $\Pi(1-\frac{z}{e^n})$ all are E-pseudo-prime functions.

THEOREM 2.8 [33,III]. Let F be an entire function of finite order for which $F'(z) = 0$ has only real zeros. Then F is E-pseudo-prime.

Similar results for meromorphic functions were obtained in [33,IV], [33,V].

THEOREM 2.9 [35,I]. Let $F(z) = \Pi(1-\frac{z}{a_\ell})^{p_\ell}$; $a_\ell > 0$, $a_{\ell+1} > a$ be an entire function of order less than one. Suppose that there are two indices j and k such that $(p_j, p_k) = 1$. Further suppose that there is a sequence $\{r_n\}$

such that $a_{n-1} < r_n < a_n$ and $\lim_{n\to\infty} F(r_n) = +\infty$. Then F is E-prime.

Example. $\prod_{n=1}^{\infty}(1-\frac{z}{\alpha_n})$, $\alpha_n > 0$, $\alpha_{n+1} \geqq k\alpha_n$, $k > 1$ and $\prod_{n=1}^{\infty}(1+\frac{z}{n^{\alpha}})$, $\alpha > 2$ are E-prime functions.

It was conjectured in [35,I] that $\prod_{n=1}^{\infty}(1+\frac{z}{n^{\alpha}})$, $2 \geqq \alpha > 1$ and $\frac{1}{\Gamma(z)}$ are E-prime functions and were proved later by Gross and Yang [20].

THEOREM 2.10 [35,I]. Let L(z) be an entire function satisfying the condition in Theorem 2.9 with the same notations. Further

$$\varliminf_{n\to\infty} \frac{\log\log |L(r_n)|}{\log r_n} = \rho,\ 0 < \rho < 1.$$

Let M(z) be an entire function having only negative zeros and being of order less than ρ. Then $F(z) = L(z)M(z)$ is E-prime.

THEOREM 2.11 [35,I]. Suppose that $\exp(H(z))$ is of hyperorder less than one, where the hyperorder of f stands for $\varlimsup_{r\to\infty} \frac{\log\log T(r,f)}{\log r}$. Then ze^H is prime.

THEOREM 2.12 [35,II]. Let F(z) be an entire function of order ρ, $\frac{1}{2} < \rho < 1$, and with only negative real zeros. Assume that n(r) (the counting function of the zeros of F) $\sim \lambda r^{\rho}$, $\lambda > 0$. Further assume that there are two indices j and k such that a_j, a_k are zeros of F(z) where multiplicities p_j, p_k satisfying $(p_j, p_k) = 1$. Then F(z) is prime.

The function $\cos\sqrt{z} = 2\cos^2\frac{\sqrt{z}}{2} - 1$ shows that if the order is not greater than $\frac{1}{2}$ the assertion does not remain true. For entire functions of order $> \frac{1}{2}$ the following result was proved by Ozawa [35,II] and Kimura [25].

THEOREM 2.13. Let F(z) be an entire function of non-integral order $\rho(\frac{1}{2} < \rho < \infty)$, the zeros of which form a simple set lying on the negative real axis. If $n(r) \sim \lambda r^{\rho}$, $\lambda > 0$, then F(z) is prime.

By weakening the condition imposed on the above function F(z), Prokopovich obtained the follwing two general results:

THEOREM 2.14 [46]. Let F(z) be a meromorphic function of finite order and suppose that a is a complex number finite or infinite which is not a Picard exceptional value of F(z). Assume that almost all a-points of F(z) lie on some L. Then F(z) is pseudo-prime. Furthermore, if $F(z) = f(g(z))$ with f being transcendental, then g(z) must be a polynomial of at most second degree.

THEOREM 2.15 [46]. Let F(z) be a meromorphic function of order $\rho(\frac{1}{2} < \rho < \infty)$, $\ell_0 = \{z: \arg z = \phi_0\}$, $\ell_\infty = \{z: \arg z = \phi_0 + \theta\}$, where $0 \leqq \theta < \min\{\pi/\rho, 2\pi-\pi/\rho\}$. Let the zeros of F(z) form an infinite P-multiple set and let almost all of them lie on ℓ_0. In addition suppose that the poles form a q-multiple set and that almost all of the lie on ℓ_0 ℓ_∞. If for some point $a \in \bar{C}$ (the extended plane) $\delta(a,F) = 1$, then F(z) is a pseudo-prime function and in the factorization $F(z) = f(g(z))$, f(z) must assume one of the following forms:

$$f(z) = A(z - \alpha_1)^m ,$$

$$f(z) = A(z - \alpha_2)^{-n} ,$$

$$f(z) = A\frac{(z - \alpha_1)^m}{(z - \alpha_2)^n} ,$$

where m divides p, n divides q, and A, α_1 and α_2 are constants. Moreover, in the last case $\delta(\alpha_1,g) + \delta(\alpha_2,g) = 1$.

It was indicated in [46] by examples: $\cos\sqrt{2}z = (\sqrt{2}\cos\sqrt{z} + 1)(\sqrt{2}\cos\sqrt{z} - 1)$, $\cos\sqrt{z}\, e^{\cos\sqrt{z}}$ and $(1 + z^2)e^{1+z^2}$ that neither the condition $\frac{1}{2} < \rho < \infty$, nor the condition on the existence of an infinite number of zeros of F(z) can be exempted.

3. PRIME FUNCTIONS IN J(2πi) or L(2πi)

In [52] the following results were obtained.

THEOREM 3.1 [52, Theorem 9]. Let $F(z) = z + h_1(e^z) + h_2(e^z)e^{e^z}$, where h_1 and h_2 ($\neq 0$) are entire functions with $\rho(h_j) < 1$ $(j = 1,2)$. Then F(z) is prime, unless h_1 is a linear polynomial and $h_2(z) = cz^m$, for some nonzero constant c and some positive integer m.

THEOREM 3.2 [52, Theorem 10]. Let $F(z) = z + Q(e^z) + h(e^{e^z})$, where h ($\neq$ constant) is entire with $\rho(h(e^z)) < \infty$ and Q is a polynomial, then F(z) is prime.

THEOREM 3.3 [52, Theorem 11]. Let $F(z) = z + h(e^z(+ Q(e^{e^z}))$, where h is entire with $\rho(h) < 1$ and Q is a nonconstant polynomial, then F(z) is prime.

THEOREM 3.4 [52, Theorem 12]. Let $F(z) = Q(e^z) + z \exp z + P(z)$ with polynomials P and Q, then F(z) is prime.

THEOREM 3.5 [52, Theorem 13]. Let $F(z) = (z + H_1(z))e^{H_2(z)}$ where H_i ($\neq$ constant) is entire, periodic with period $2\pi i$ and $\rho(H_i) < \infty$ $(i = 1,2)$, then F(z) is prime.

4. FUNCTIONS AND THEIR DERIVATIVES ALL ARE E-PRIMENESS

THEOREM 4. [56]. Let $F(z) = \int_0^z (e^t-1)e^{t^2}\,dt$. Then $F^{(n)}$ is E-prime for any $n = 0, 1, 2, \ldots$

K. Niino proved the following result.

THEOREM 4.1 [31]. Let $F(z) = \int_0^z (e^t-1)e^{t^k}\,dt$ $(k \geqq 3$; an integer), then $F^{(n)}(z)$ is E-prime for $n = 0, 1, 2, \ldots$.

In the same paper Niino also made a study of the factorization of $F^{(n)}(z)$, where F(z) is a function of the form

$$F(z) = \int_0^z (H_1(t)e^t + H_2(t))e^{P(t)}\,dt \quad ,$$

and $H_i(z)$ $(\not\equiv 0)$ $(i = 1, 2)$ are entire functions of order less than one with the zeros of $H_1(z)e^z + H_2(z)$ satisfying certain conditions and P(z) is a polynomial of degree not lower than two.

THEOREM 4.2 [31]. Let P(z) be a polynomial of degree k $(k \geqq 3)$ such that P(z) is arbitrary if k is odd and $P(z) = \alpha_k z^k$ $(\alpha_k \neq 0)$ if k is even. Let $P_1(z)$ and $P_2(z)$ be two polynomials which are not identically zero. Then all $F^{(n)}(z)$, $n = 0, 1, 2, \ldots$ are prime, where

$$F(z) = \int_0^z \{P_1(t)e^t + P_2(t)\}\, e^{P(t)}\,dt \quad .$$

Remarks:

(i) As exhibited in [31]the above result may not be true in the case when k = 2.

(ii) The condition on the zeros of $H_1(z)e^z + H_2(z)$ can be removed [32].

5. PRIME FUNCTIONS WITH GAP POWER SERIES

By method based on results of Erdős and Macintyre, Liverpool [27] proved the following interesting result:

THEOREM 5.1. Let $\{p_n\}$ be a sequence of primes such that

$$\sum_{n=0}^{\infty} \frac{1}{p_{n+1}-p_n} < \infty .$$

Let $F(z) = \sum_{n=0}^{\infty} a_n z^{p_n}$ be an entire function. Then F(z) is prime.

In [17, p. 248] the following question was raised and is still an open one. Question: Let f(z) be any entire function of the form $f(z) = \sum_{n=0}^{\infty} a_n z^{p_n}$ where p_n are all primes. Is f(z) necessarily prime?

6. FACTORS OF PERIODIC AND ELLIPTIC FUNCTIONS

We have seen that, in general, the factors of periodic entire functions are quite restricted. Some additional properties of factors of periodic and elliptic functions will be stated below.

THEOREM 6.1 [47]. Let p be a polynomial and g be meromorphic, then p(g) is periodic if and only if g is.

Gross [11] generalized the above result as follows:

THEOREM 6.2. Let f be entire of order less than ½, and g be meromorphic, then f(g) is periodic if and only if g is.

It was remarked in [11], f can be replaced by a different class of functions as in the following statement.

THEOREM 6.3. Let f be a meromorphic function having at most a finite number of fix-points. Then f(g) is periodic if and only if g is.

We thus have realized that for periodic entire functions when their left factors belong to a certain class of meromorphic functions, all other entire right factors are ruled out except periodic ones. Furthermore, there are certain classes of right entire factors of periodic functions are ruled out without any restrictions posed on the left factors. The following result was obtained by I. N. Baker [3] and A. and C. Renyi [47] independently.

THEOREM 6.4. Let p be a polynomial of degree greater than 2 and if f is transcendental entire, then F = f(p) is not periodic.

In 1963, in a private communication to Gross, A. Renyi conjectured that the above result remains valid when p is replaced by any entire function of order less than 1. This conjecture was then proved by G. Halasz in 1972 [22] as follows.

THEOREM 6.5. Let f be entire and g entire of order less than 1. If f(g) is periodic, then g must be a polynomial of degree 2.

For meromorphic factors Fuchs and Gross [7] proved the following result analogous to Theorem 6.4.

THEOREM 6.6. Let f be nonconstant meromorphic function and g be a polynomial of degree n. The function F(z) = f(p)(z) cannot be periodic unless n has the values 1, 2, 3, 4 or 6.

Conjecture [16]. Theorem 6.6 remains valid when p is replaced by any entire function of order less than 1.

Concerning factors of elliptic functions the following results were obtained by Gross [13].

THEOREM 6.7. Let h be an elliptic function with left and right factors f and g respectively. If g is entire, then g has no defect values.

THEOREM 6.8. An elliptic function cannot have a periodic left factor.

THEOREM 6.9. A right factor of an elliptic function of valence 2 which is not itself elliptic must be either a polynomial of degree 2 or of the form a cos (cz + b) + d, where a, b, c, d are constants.

THEOREM 6.10. Let $P(z)$ be double periodic Weierstrass function, then $P(z)$ is pseudo-prime, and its only possible nonelliptic right factors are cubic polynomials.

THEOREM 6.11. A transcendental right factor of an elliptic function must be periodic if it is entire and elliptic otherwise.

THEOREM 6.12. An elliptic function $\phi(z)$ has a common right factor with a co-periodic elliptic function $\psi(z)$ of valence 2 if and only if ϕ is a rational function of ψ.

7. ON UNIQUE FACTORIZATION

It is natural for one to consider the uniquely factorizable question. If every factorization of $F(z)$ into nonlinear prime entire factors is equivalent to one another, then we say that $F(z)$ is uniquely factorizable. This subject opens a new branch in the development of the factorization theory. e^z and $\sin z$ are pseudo-prime and admit infinitely many nonequivalent factorizations. If one follows [52] and takes $F(z) = z^p \circ \exp(z^p)$ with a prime p (≥ 2), then $F(z)$ has two nonequivalent factorization of prime factors; $z^p \circ ze^{z^p/p} = ze^z \circ z^p$. The first studied uniquely factorizable function seems to be z^2e^{2z} which was showed in [17, p. 133]. It was pointed in [37] that, in general, $F(z) = z^pe^{pz} = z^p \circ (ze^z)$ is uniquely factorizable if p is a prime number and further $F(z) = p(z)e^{p(z)} = (ze^z) \circ p(z)$ is so if $p(z)$ is a nonlinear polynomial which is prime and has at least one simple zero or two zeros with coprime multiplicities. The function $F(z) = (ze^z) \circ (ze^z)$ seems to be the simplest example which is uniquely factorizable and with both factors are transcendental [52]. More is true.

THEOREM 7.1 [52, p. 105]. Let p and q be two nonconstant polynomials, then $(ze^z) \circ (h(z)e^z)$ and $(ze^{p(z)}) \circ (ze^{q(z)})$ are uniquely factorizable functions.

As it is well-known that $e^z + z$ is a prime function, one of the purposes of the paper [52] was to settle the question whether or not the composite function $F(z) = (z+e^z)\circ(z+e^z)$ is uniquely factorizable. The above question was answered affirmatively and contained in the following result.

THEOREM 7.2 [52, Theorem 3]. Let $F(z) = (z + h(e^z)) \circ (z + Q(e^z))$, where $h(z)$ is a nonconstant entire function with the order $\rho(h(e^z)) < \infty$ and $Q(z)$ is a nonconstant polynomial. Then $F(z)$ is uniquely factorizable.

In [52] the following four results were obtained.

THEOREM 7.3. Let $F(z) = (H(z) + z \exp [z + h(e^z)]) \circ (z + p(e^z))$, where $H(z)$ and $h(z)$ are entire functions with $H(z = 2\pi i) = H(z)$ and the order of $h(e^z) < \infty$, and $p(z)$ is a nonconstant polynomial. Assume that the function $H(z) + z \exp (z + h(e^z))$ is prime, then $F(z)$ is uniquely factorizable.

THEOREM 7.4. Let $F(z) = (H_1(z) + ze^z) \circ (z + H_2(z))$, where entire functions $H_i (i = 1,2)$ have period $2\pi i$. Assume that $z + H_2(z)$ is prime, then $F(z)$ is uniquely factorizable.

THEOREM 7.5. Let $F(z) = (H_1(z) + ze_m(z)) \circ (z + H_2(z))$, where H_j $(j = 1, 2)$ are periodic entire functions with period $2\pi i$ and the order of $H_2 < \infty$. Then $F(z)$ is uniquely factorizable.

THEOREM 7.6. Let $F(z) = (z + H_1(z))e^{H_2(z)}$, where entire functions H_1 and e^{H_2} have period $2\pi i$ with the order of $H_1 < \infty$. Then $F(z)$ is uniquely factorizable.

Further Theorem 7.6 can be generalized as follows:

THEOREM 7.7 [54]. Let $F(z) = (z + h_1(z)) \exp z + H_2(z)$, where $H_j(z)$ $(j = 1, 2)$ are entire, periodic, with period $2\pi i$. Then $F(z)$ has a nontrivial factorization if and only if $z + H_1(z)$ and $z + H_2(z)$ have the common, nonlinear entire, right factor (which necessarily belongs to $J(2\pi i)$).

For other types of uniquely factorizable entire functions we refer the reader to Ozawa [37]. The following result is contained therein.

THEOREM 7.8 [37, Theorem 3]. Let $\{a_n\}$ be a set of complex numbers and $\{v_n\}$ be a set of positive integers satisfying $v_1 < v_2 < \dots < v_n < v_{n+1} < \dots$. Assume that $\sum_{n=1}^{\infty} \frac{v_n}{a_n^s} < \infty$, for $s < 1$. Further assume that v_1 and v_2 are coprime. let $g_1(z)$ be $\sum_{n=1}^{\infty} (1-\frac{z}{a_n})v_n$. Then $\{g_1(z)\}^2$, $g_1(z)\exp q(g_1(z))$ are uniquely factorizable into two primes, where $q(z)$ is a polynomial.

Recently Ozawa also dealt with uniquely factorizable of certain meromorphic functions [44].

We conclude this paper by mentioning the following significant result in the factorization theory due to Ozawa [40]:

THEOREM 7.9. Let F be an entire function, $m = 3, 2^j, j = 1, 2, \ldots$. Suppose that for each m, there exist polynomial P_m of degree m and entire function f_m such that $F(z) \equiv P_m(f_m(z))$. Then $F(z) \equiv Ae^{H(z)} + B$ or $F(z) \equiv A \cos \sqrt{H(z)} + B$; A, B constants and H entire.

Subsequent to this a series of papers dealt with the characterization of exponential functions and cosine functions have been written by Ozawa [41, 42, 43].

REFERENCES

1. I. N. Baker, The iteration of entire transcendental functions and the solution of the functional equation f(f(z))= F(z), *Math. Ann.* 129 (1955), 174-180.

2. I. N. Baker, Permutable entire functions, *Math. Zeit* 79, (1962) 243-249.

3. I. N. Baker, Some results of A. Renyi and C. Renyi concerning periodic entire functions, *Acta Sci. Math.* 27, Nos. 3-4 (1966), 197-200.

4. I. N. Baker and F. Gross, Further results on factorization of entire functions, Proc. Symposia Pure Math. (*Amer. Math. Soc.,* Providence, R.I.) II (1968), 30-35.

5. I. N. Baker and C. C. Yang, An inifinite order periodic entire function which is prime, Lecture Notes in Mathematics, Vol. 599, Complex Analysis - Kentucky 1976, Springer Verlag 7-10.

6. P. Fatou, Sur literation des fonctions transcendants entières, *Acta Math.* 47 (1926), 337-370.

7. W. H. J. Fuchs and F. Gross, Generalization of a theorem of A. and C. Renyi on periodic functions, *Acta Sci. Math.,* 32. Mps/ ;=2 (1971), 83-86.

8. A. A. Gol'dberg and G. S. Prokopovich, On the simplicity of certain entire functions, *Ukrainskii Matematicheskii Zhurnal*, Vol. 22, No. 6 (1970), 813-817. (English translation 701-704.)

9. R. Goldstein, On factorization of certain entire functions, *L. London Math. Soc.* (2) (1970), 221-224.

10. R. Goldstein, On factorization of certain entire functions, II, Proc. London Math. Soc. Vol. 22 (1971), 483-506.

11. F. Gross, On the distribution of values of meromorphic functions, *Trans. Amer. Math. Soc.* Vol. 131 (1968), 199-214.

12. F. Gross, On the periodicity of compositions of entire functions, *Canada J. Math.* 18, No. 4 (1966), 724-730.

13. F. Gross, On factorization of elliptic functions, *Canadian J. Math.* (20) (1968), 486-494.

14. F. Gross, On factorization of meromorphic functions, *Trans. Amer. Math. Soc.*, Vol. 131 (1968), 215-222.

15. F. Gross, Prime entire functions, *Trans. Amer. Math. Soc.* 161 (1971), 219-233.

16. F. Gross, Factorization of meromorphic functions and some open questions, complex analysis, Lecture Notes, Springer-Verlag, N.Y., 1976.

17. F. Gross, Factorization of meromorphic function, U.S. Government Printing Office, Washington, D.C., 1973.

18. F. Gross and C. C. Yang, The fix points and factorization of meromorphic functions, *Trans. Amer. Math. Soc.* 168 (1972) 211-219.

19. F. Gross and C. C. Yang, Further results on prime entire functions, *Trans. Amer. Math. Soc.*, Vol. 142 (1974), 347-355.

20. F. Gross and C. C. Yang, On factorization of entire functions with infinitely many real zeros, *Ind. J. Pure and App. Math.* 3 (1972), 1183-1194.

21. W. K. Hayman, Meromorphic function, *Oxford Mathematical Monographs,* Clarendon Press, Oxford, 1964.

22. G. Halasz, On periodicity of composed integral functions, *Period Math. Hunger* 2, (1972), 78-83.

23. G. Julia, Mémoires sur literation des fonctions rationelles, *J. Math.* (7), 4 (renumbered (8), 1) (1918), 47-245.

24. G. Julia, Memoires sur les permutabilite des fonctions rationelles, *Ann. Sci. de l'Ecole Normale Superieure* (3) 39, (1922), 131-215.

25. S. Kimura, On prime entire functions, *Kodai Math. Sem. Rep.* 24 (1972), 28-33.

26. S. Koont, On factorization in certain classes of entire functions, Math. Research report, No. 76-5, March 1976, University of Maryland - Baltimore County, MD, U.S.A.

27. L. S. O. Liverpool, Some remarks on factorization of entire functions, *J. London Math. Soc.* (2) 7 (1973).

28. E. Mues, "Uber factor; sierbare Losungen von Riccatischen Differentialgleichungen," *Math. Zeit.* 121 (1971) 145-156.

29. E. Mues, Zur Faktorisierung elliptischer Funktionen, *Math. Zeit.* 120, No. 2 (1971), 157-164.

30. K. Niino, On factorization of certain entire functions, *Kodai Math. J.* 1 (1978), 277-284.

31. K. Niino, On prime entire functions, *J. Math. Anal. Appl.* 66 (1978), 178-187.

32. K. Niino and C. -C. Yang, On a class of prime entire functions, to appear.

33. M. Ozawa, On the solution of the functional equation fog(z) = F(z), I, II, III, IV and V, *Kodai Math. Sem.·Rep.* (20) (1968), 159-162, 163-169, 257-263, 272-278 and 305-313, respectively.

34. M. Ozawa, Factorization of entire functions, *Tohoku Math. Journal* (27), (1975), 321-336.

35. M. Ozawa, On prime entire functions, I and II, *Kodai Math. Sem. Rep.* (22), (1975), 301-308 and 309-312, respectively.

36. M. Ozawa, On certain criteria for the left-primeness of entire functions, *Kodai Math. Sem. Rep.* 26 (1975), 304-317.

37. M. Ozawa, On uniquely factorizable entire functions, *Kodai Math. Sem. Rep.* 8 (1977), 342-360.

38. M. Ozawa, Sufficient conditions for an entire function to be pseudo-prime, *Kodai Math. Sem. Rep.* 27 (1976), 373-378.

39. M. Ozawa, On the existence of periodic entire prime functions, *Kodai Math. Sem. Rep.* 29 (1978), 308-321.

40. M. Ozawa, On a characterization of the exponent function and the cosine function by factorization, *Kodai Math. Journal* (1978), 45-74.

41. M. Ozawa, On a characterization of the exponential function and the cosine function by factorization; II, III *Kodai Math. Jour.* 1 (1978), 313-315, 1 (1978), 200-210.

42. M. Ozawa, A characterization of the exponential function and the cosine function by factorization, IV, This Notes.

43. M. Ozawa, A characterization of the cosine function by the value distribution, *Kodai Math. Jour.* 1 (1978), 213-218.

44. M. Ozawa, On uniquely factorizable meromorphic function, *Kodai Math. Jour.* 1 (1978), 339-353.

45. G. S. Prokovich, On superposition of some entire functions, *Ukrainskii Matematicheskii Zhurnal*, Vol. 26, No. 2, March-April, 1974, 188-195.

46. G. S. Prokovich, On pseudo-simplicity of some meromorphic functions, *Ukrainskii Matematicheskii Zhurnal,* Vol. 27, No. 2, March-April 1975, 261-273. (English transl. 219-222.)

47. A. Renyi and C. Renyi, Some remarks on periodic entire functions, *J. Analyse Math.* 14 (1965), 303-310.

48. J. F. Ritt, Prime and composite polynomials, *Trans. Amer. Math. Soc.* 23 (1922), 51-66.

49. P. C. Rosenbloom, The fix-points of entire functions, Medd Lunds Univ. Mat. Sem. Suppl. Bd., M. Riesz (1952), 186-192.

50. N. Steinmetz, On factorization of the solutions of the Schwarzian differential equation {w,z} = q(z), to appear.

51. N. Steinmetz, Uber die faktorisierbaren Lösungen gewöhnlicher Differentialgleichungen, *Math. Zeit.* 170 (1980), 169-180.

52. H. Urabe, Uniqueness of the factorization under composition of certain entire functions, *Jour. of Math. of Kyoto Univ.*, Vol. 18, No. 1 (1978), 95-120.

53. H. Urabe, On a factorization problem of certain entire functions and a theorem of Ostrovskii, Bulletin of Kyoto Univ. of Education, Ser. B, No. 53, 1978.

54. H. Urabe, Private communication.

55. H. Urabe, Infinite order periodic, entire functions which together with their derivatives are prime, Bulletin of the Kyoto Univ. of Education, Ser. B, No. 55, 1979.

56. H. Urabe and C. C. Yang, On certain entire functions which together with their derivatives are prime, *Kodai Math. Sem. Rep.* 29 (1977), 167-178.

57. C. C. Yang, On the factorization of entire function, *Ill. Journal of Math.*, Vol. 21, No. 4 (1977), 898-905.

INDEX